Green Energy and Technology

Ayhan Demirbas

Methane Gas Hydrate

Springer

Ayhan Demirbas, Professor of Energy Technology
Sila Science and Energy
Trabzon
Turkey

ISSN 1865-3529 e-ISSN 1865-3537
ISBN 978-1-4471-2541-9 e-ISBN 978-1-84882-872-8
DOI 10.1007/978-1-84882-872-8
Springer London Dordrecht Heidelberg New York

British Library Cataloguing in Publication Data
A catalogue record for this book is available from the British Library

Cover design: WMXDesign, Heidelberg, Germany

Printed on acid-free paper

Springer is part of Springer Science+Business Media (www.springer.com)

Preface

Currently, natural gas is the fastest-growing primary energy source in the world. Natural gas is widely available and it is the cleanest-burning petroleum-based fuel. Natural gas can be produced from natural gas hydrates, or methane hydrates. The presence of gas hydrates in oceanic sediments was first postulated on the basis of seismic observations. Gas hydrates represent one of the world's largest untapped reservoirs of energy and, according to some estimates, have the potential to meet global energy needs for the next 1,000 years. They are also of considerable interest because of their potential role in global climate change. From an energy resource point of view, the enormous amounts of gas hydrates under the ocean and beneath Arctic permafrost represent an estimated more than 50% of all carbonaceous fuel reserves on Earth. The significance of methane hydrates lies mainly in the tremendous potential of this resource. Even a small percentage of the estimated global resource, if exploited, could meet the world's energy demands for centuries.

The industrial and social changes in which mankind was engaged during the twentieth century resulted in a rapidly increasing demand for hydrocarbon-based fuels. This increasing demand has been met by increasing the speed of exploration and exploitation of the vast resources of conventional oil and gas. These resources, however, are neither renewable nor inexhaustible, and we should expect to encounter severe depletion in the not-too-distant future. In this scenario, natural gas hydrates appear to be an exciting alternative and, if approached holistically, may become the main and a bountiful source of fuel for an energy-hungry world.

Gas hydrates are crystalline solid compounds, consisting of a gas molecule surrounded by a cage of water molecules. They are stable under the high pressure and low temperature typical of deepwater sediments of the world's oceans. Gas hydrate, or methane hydrate, is composed of natural gas molecules trapped inside ice. Per unit mass, gas hydrates contain twice as much carbon as all other fossil fuels (coal, natural gas, and oil) combined. The amount of methane trapped in marine sediments as a hydrate represents such an immense carbon reservoir that it must be considered a dominant factor in estimating unconventional gas energy resources. Significant safety and environmental concerns are also associated with the presence of natural gas hydrates, ranging from their possible impact on the safety of conventional drilling operations to the influence on Earth's periodic natural releases into the atmosphere of large volumes of hydrate-sourced methane or its derivative carbon dioxide.

Three processes have been proposed for dissociation of methane hydrates: thermal stimulation, depressurization, and inhibitor injection. The obvious production approaches involve depressurization, heating, and their combinations. The depressurization method involves lowering the pressure inside the well and encouraging the methane hydrate to dissociate. The chemical inhibition method seeks to displace the natural gas hydrate equilibrium condition beyond the hydrate stability zone's thermodynamic conditions through injection of a liquid inhibitor chemical adjacent to the hydrate. Of these three production processes, depressurization combined with the thermal stimulation process appears to be the most practical for zones where free gas is trapped beneath the methane hydrates.

Methane Gas Hydrate is written for the field engineer working in the natural gas industry. This book explains how, when, and where hydrates form, and provides the knowledge necessary to apply remedies in practical applications.

This book attempts to address the needs of energy researchers, chemical engineers, chemical engineering students, geology engineers, geology engineering students, geophysicists, gas engineers, energy resources specialists, engineers, and others interested in a practical tool for pursuing their interests in relation to energy. Each chapter starts with basic/fundamental explanations suitable for general readers and ends with in-depth scientific details suitable for expert readers. The expert readers will include chemists, chemical engineers, fuel engineers, geology engineers, geophysics engineers, geophysicists, gas engineers, biologists, fuel processors, policymakers, environmentalists, environmental engineers, automobile engineers, college students, and research faculty.

Contents

Chapter 1
Introduction

1.1 Global Energy Sources

Global energy sources are classified into two groups: fossil and renewable. Primary energy sources can be divided into nonrenewables and renewables. Nonrenewable energy resources include coal, petrol, gas, gas hydrate, and fissile material, whereas biomass, hydro, geothermal, solar, and wind energy are renewable energy sources. The energy source can also be classified according to its depletion rate. Whereas biomass can be depleted, the sun and the wind are nondepletable. Reality shows that the energy available from nonrenewable sources is limited, and that the exploration, the processing, and the use of energy have considerable impacts on the environment.

Fossil fuels have been the prime sources of energy for the purpose of transportation, power generation, agriculture, and commercial, domestic, and industrial activity for more than a century. The world's energy requirements are currently satisfied by fossil fuels, which serve as the primary energy source. Overwhelming scientific evidence has shown that this unfettered use of fossil fuels has caused the world's climate to change, with potentially disastrous effects. Moreover, the oil crisis in 1973 provided a reminder that breaking an energy paradigm based on fossil fuel dependency would lead to economic and environmental advantages.

Because of the increase in petroleum prices, especially after the oil crisis in 1973 and the Gulf War in 1991, geographically reduced availability of petroleum, and the imposition of more stringent governmental regulations on exhaust emissions, researchers have studied alternative fuels and alternative solution methods (Durgun and Sahin 2007). The term "fossil" refers to an earlier geological age. Fossil fuels were formed a great many years ago and are not renewable. The fossil energy sources are petroleum (crude oil), coal, bitumen, natural gas, oil shales, and tar sands. During the last 200 years, developed countries have shifted their energy consumption toward fossil fuels. Petroleum is the largest single source of energy consumed by the world's population, exceeding coal, natural gas, nuclear power,

A. Demirbas, *Methane Gas Hydrate*,

hydropower, and renewables. About 98% of carbon emissions result from fossil fuel combustion. Reducing the use of fossil fuels would considerably reduce the amount of carbon dioxide and other pollutants produced. This can be achieved either by using less energy altogether or by replacing fossil fuels by renewable fuels. Hence, current efforts focus on advancing technologies that emit less carbon or no carbon, such as nuclear, hydrogen, solar, wind, and geothermal technologies, or on using energy more efficiently and on developing sequestration of the carbon dioxide emitted during fossil fuel combustion. In fact today, over 80% of the energy we use comes from three fossil fuels: petroleum, coal, and natural gas. Unfortunately, petroleum oil is in danger of becoming in short supply. Hence, the future trend is toward using alternative energy sources. Fortunately, technological developments are making the transition possible.

Another problem with petroleum fuels is their uneven distribution in the world; for example, the Middle East has 63% of the global reserves and is the dominant supplier of petroleum. This energy system is unsustainable because of equity issues as well as environmental, economic, and geopolitical concerns that have far-reaching implications. Interestingly, the renewable energy resources are more evenly distributed than fossil or nuclear resources. Hence, sustainable renewable energy sources such as biomass, hydro, wind, solar (both thermal and photovoltaic), geothermal, and marine energy sources will play an important role in the world's future energy supply.

Developing renewable sources of energy has become necessary owing to the limited supply of fossil fuels. Global environmental concerns and decreasing resources of crude oil have prompted demand for alternative fuels. Global climate change is also the major environmental issue of current times. Global warming, the Kyoto Protocol, the emission of greenhouse gases, and the depletion of fossil fuels are the topics of environmental pleadings worldwide. Rapidly increasing energy requirements parallel technological developments, and research and development activities are forced to study new and renewable energy sources.

The major sources of alternative energy each have their own advantages and disadvantages, including political, economic, and practical issues. Renewable energy is a promising alternative solution because it is clean and environmentally safe. Renewable energy sources also produce lower or negligible levels of greenhouse gases and other pollutants when compared with the fossil energy sources they are replacing.

Competition between renewable liquid fuels and petroleum liquid fuels will be important in the near future. This can be achieved by research and development, technological development, and industrial mobilization, by implementing a proper energy tax system which takes into account the environmental and social damage caused by conventional fuels, and by making the right calculations for the whole energy system, and not only calculating the cost of one technology but the cost of all of the energy chain, to calculate the burdens for every foreign national economy as a result of importing fossil energies, etc. This needs a comprehensive view on energy.

1.2 Fossil Fuel Sources

Fossil fuels were formed a great many years ago and are not renewable. The fossil energy sources are petroleum (crude oil), coal, bitumen, natural gas, oil shales, and tar sands. Figure 1.1 shows the fossil energy reserves of the world. The world's coal reserves correspond to 3.2×10^{17} MJ. On the other hand, the world's deuterium and uranium reserves correspond to 7.5×10^{24} and 1.2×10^{20} MJ, respectively (Demirbas 2009). Petroleum is the largest single source of energy consumed by the world's population, exceeding coal, natural gas, nuclear power, hydropower, and renewables, as shown in Fig. 1.2 for the year 2005. About 98% of carbon emissions result from fossil fuel combustion. Reducing the use of fossil fuels would considerably reduce the amount of carbon dioxide and other pollutants produced. This can be achieved either by using less energy altogether or by replacing fossil fuels by renewable fuels.

Methane hydrates are the world's largest source of fossil fuel. Large fields of methane hydrates are scattered throughout the world's oceans and are thought to contain about as much energy as all other forms of fossil fuel combined. This unconventional hydrocarbon energy source has remained untapped, however, because traditional sources are still plentiful and less expensive to develop. Nevertheless, scientists have recently been taking a closer look at hydrates, and not only as a possible energy source. A widely quoted US Geological Survey estimate predicts that there is twice as much organic carbon in gas hydrate than in all recoverable and unrecoverable conventional fossil fuel resources, including natural gas, coal, and oil.

Before a coal population ceiling was reached, a new source of energy replaced coal's dominance. Oil was the next source of energy to be commercialized. Oil is easier to handle than coal. It is burns more cleanly and is cheaper to transport and

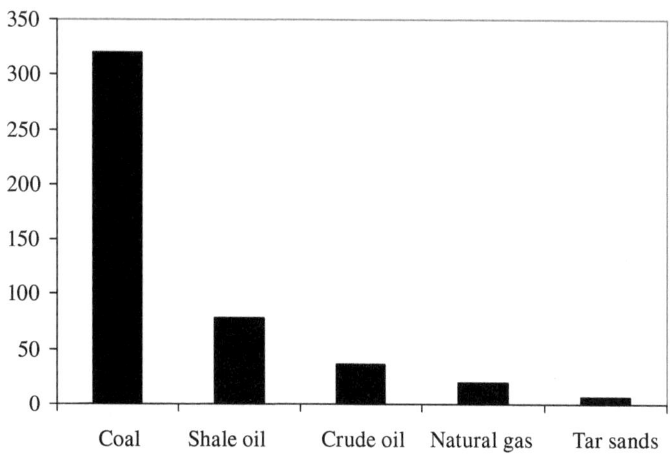

Fig. 1.1 Fossil energy reserves of the world. Each unit corresponds to 1×10^{15} MJ, equivalent to 1.67×10^{11} billion barrels of crude oil

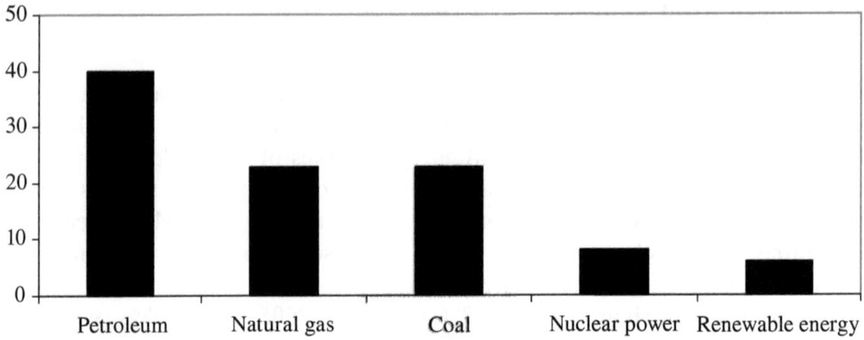

Fig. 1.2 World's energy consumption in 2005

store, making it ideal as a transportation fuel. It has a higher thermodynamic potential than coal, and its use was able to further increase productivity and arguably led to less land-use demand. It is clear that the availability of fossil fuels, in particular crude oil, has had a profound effect on population growth. The population has grown because death rates have declined worldwide, but birth rates have remained at high levels in many parts of the world. Oil arguably plays a part in both phenomena.

Oil also plays a significant part in the so-called green revolution that has led to growth in agricultural output that has managed to keep up with or even exceed the number of mouths that require feeding. Green revolution agriculture relies on large amounts of pesticides and fertilizers, products highly dependent on oil and gas. Intensification of agriculture leads to surplus production, enabling greater increases in population, which in turn leads to still greater demands for food.

1.2.1 The History of Petroleum

Oil is the fossil fuel that is most in danger of becoming in short supply. The Middle East is the dominant oil region of the world, covering 63% of the global reserves. Countries in the Middle East and the Russian Federation hold 70% of the world's dwindling reserves of oil and gas. The geographic distribution of energy reserves and resources is important.

The modern history of petroleum began in 1846 with the discovery of the refining of kerosene from coal by Atlantic Canada's Abraham Pineo Gesner. Poland's Ignacy Łukasiewicz discovered a means of refining kerosene from the more readily available "rock oil" ("petroleum") in 1852, and in the following year the first rock oil mine was built in Bobrka, near Krosno in southern Poland. News of the discovery rapidly spread around the world, and Meerzoeff built the first Russian refinery in the mature oil fields of Baku in 1861, which produced about 90% of the world's oil. In fact, the battle of Stalingrad was fought over Baku.

The first commercial crude oil well in North America was drilled by James Miller Williams in 1858 in Oil Springs, Ontario, Canada. In the following year, Edwin Drake discovered oil near Titusville, Pennsylvania, USA, and pioneered a new method for producing oil from the ground, in which he drilled using piping to prevent borehole collapse, allowing for the drill to penetrate deeper into the ground.

For about 10 years, Pennsylvania was the one great crude oil producers of the world, but since 1870 the industry has spread all over the globe. Since the time of the completion of the first flowing well on the Baku field, Russia has ranked second on the list of producing countries, whereas Galicia and Romania became prominent in 1878 and 1880, respectively. Sumatra, Java, Myanmar, and Borneo, where active development began in 1883, 1886, 1890, and 1896, respectively, will before long rank before long among the chief sources of world oil supplies.

Before the 1850s, Americans often used whale oil to light their homes and businesses. Drake refined the crude oil from his well into kerosene for lighting, which was used till the discovery of light bulbs. Gasoline and other products made during refining were simply discarded owing to lack of use. In 1892, the "horseless carriage" solved this problem since it required gasoline. By 1920 there were nine million motor vehicles in the USA and many gas stations to supply gasoline.

In 1858, Dr. Abraham Gesner, a Canadian physician and amateur geologist, developed and patented the extraction of a lamp fuel from asphalt rock, which he named kerosene. At that time, kerosene, which was an extremely high value lighting fuel for lamps, was the primary product of petroleum refining. For a while, distillation of kerosene for lamps was the mainstay of the new petroleum industry. Gasoline was merely a by-product of kerosene production from crude oil and, until the early 1900s, there was no significant demand for it. The first petrochemical, aside from carbon black manufactured on an industrial scale, was isopropyl alcohol, produced by Standard Oil of New Jersey in 1920.

There are two great petroleum organizations in the world: the Organization of Petroleum Exporting Countries (OPEC) and the Organization for Economic Cooperation and Development (OECD). Iran, Iraq, Kuwait, Saudi Arabia, Venezuela, Qatar, Indonesia, Libya, the United Arab Emirates, Algeria, Nigeria, Ecuador, and Gabon are currently members of OPEC. The OPEC members try to set production levels for petroleum to maximize their revenue. According to supply/demand economics, the less oil they produce, the higher the price of oil on the world market, and the more oil they produce, the lower the price. However, the OPEC members do not always agree with each other. Some OPEC members want to produce less oil to raise prices, whereas other OPEC members want to flood the market with petroleum to reap immediate returns. In addition, the oil supply can be controlled for political reasons. For example, the 1973 OPEC oil embargo was a political statement against the USA for supporting Israel in the Yom Kippur war. Such an embargo or a cut in production causes a drastic increase in the price of petroleum.

The USA has also turned to non-Arab countries and non-OPEC members for oil imports. Today, the USA imports much of its oil from Canada and Mexico.

Table 1.1 Regional shares of crude oil production in 1973 and 2006 (IEA 2007)

Region	1973	2006
Middle East (%)	37.0	31.1
OECD (%)	23.6	23.2
Former USSR (%)	15.0	15.2
Africa (%)	10.0	12.1
Latin America (%)	8.6	9.0
Asia excluding China (%)	3.2	4.5
China (%)	1.9	4.7
Non-OECD Europe (%)	0.7	0.2
Total (million tons)	2,867	3,936

OECD Organization for Economic Cooperation and Development

This is good for the USA because it has friendly relations with its neighbors, and because the oil does not have to be shipped so far. Still, the amount of oil that the USA can import from Canada and Mexico is limited. By law, Mexico can only export half the oil it produces to the USA.

The International Energy Agency (IEA) is an international energy organization. The IEA was founded by the OECD nations after the oil shocks in the 1970s as a counterweight to OPEC. Since that time, the IEA has been regarded as the "energy watchdog" of the western world and is supposed to help avoid future crises. Until 2004 the IEA published the *World Energy Outlook* every 2 years; since then it has published it every year. The *World Energy Outlook* forecasts the development in the coming two decades. These reports are considered by many people to be something like a "bible" and are the usual basis for demand and supply forecasts. The IEA also publishes monthly reports covering the current situation of the oil markets.

Table 1.1 shows 1973 and 2006 regional shares of crude oil production.

1.2.2 Petroleum Refining

Crude oil (raw petroleum) is separated into fractions by fractional distillation. The fractions at the top are less dense than the fractions at the bottom. The heavy bottom fractions are often cracked into lighter, more useful products. All of the fractions are processed further in other refining units.

Crude oil is a complex mixture that is between 50 and 95% hydrocarbon by weight. The main crude oil components are listed in Table 1.2. An oil refinery cleans and separates the crude oil into various fuels and by-products. The most important one is gasoline. Some other petroleum products are diesel fuel, heating oil, and jet fuel. Before the actual refining begins, the stored heavy crude oil is cleaned of contaminants such as sand and water. Three major refinery processes change crude oil into finished products: (1) separation, (2) conversion, and (3) purification. The first step is to separate the crude oil into its naturally occur-

Table 1.2 Main crude oil fractions

Component	Boiling range (K)	Number of carbon atoms
Natural gas	<273	1–4
Liquefied petroleum gas	231–273	3–4
Petroleum ether	293–333	5–6
Ligroin (light naphtha)	333–373	6–7
Gasoline	313–478	5–12, and cycloalkanes
Jet fuel	378–538	8–14, and aromatics
Kerosene	423–588	10–16, and aromatics
No. 2 diesel fuel	448–638	10–22, and aromatics
Fuel oils	>548	12–70, and aromatics
Lubricating oils	>673	>20
Asphalt or petroleum coke	Nonvolatile residue	Polycyclic structures

ring components. One method is a heating process called distillation. Since oil products have different boiling points, the end products can be distilled or separated. For example, asphalts have a higher boiling point than gasolines, allowing the two to be separated.

When a mixture of two liquids of different boiling points is heated to its boiling point, the vapor contains a higher mole fraction of the liquid with the lower boiling point than the original liquid; i.e., the vapor is enriched in the more volatile component. If this vapor is now condensed, the resultant liquid is enriched in the more volatile component. This is the principle of batch fractional distillation, and in a distillation column many, many such cycles are performed continuously, allowing almost complete separation of liquid components. A generalized distillation column is shown in Fig. 1.3.

Industrial distillation is typically performed in large, vertical, steel cylindrical columns known as "distillation towers" or "distillation columns" with diameters ranging from about 65 cm to 11 m and heights ranging from about 6 to 60 m or more. To improve the separation, the tower normally has horizontal plates or trays inside, or the column is packed with a packing material. To provide the heat required for the vaporization, a reboiler is most often added to the bottom of the column.

Large-scale industrial fractionation towers use reflux to achieve more efficient separation of products. Reflux refers to the portion of the condensed overhead liquid product from a distillation tower that is returned to the upper part of the tower. Inside the tower, the downflowing reflux liquid provides cooling and partial condensation of the upflowing vapors, thereby increasing the efficacy of the distillation tower. There are generally 25–45 plates or trays in a distillation tower. Each of the plates or trays is at a different temperature and pressure. The stage at the bottom of the tower has the highest pressure and temperature. Progressing upward in the tower, the pressure and temperature decrease for each succeeding stage.

The conversion processes have focused on reducing the length of some hydrocarbon chains. The primary purpose of the conversion processes is to convert low-

Table 1.3 Common process units found in an oil refinery

Unit	Treatment
Desalter	It washes out salt from the crude oil before it enters the atmospheric distillation unit
Atmospheric distillation	It distills crude oil into fractions
Vacuum distillation	It further distills residual bottoms after atmospheric distillation
Naphtha hydrotreater	It uses hydrogen to desulfurize naphtha from atmospheric distillation
Catalytic reformer	It is used to convert the naphtha-boiling-range molecules into higher-octane reformer product (reformate)
Distillate hydrotreater	It desulfurizes distillates (such as diesel) after atmospheric distillation
Fluid catalytic cracker	It upgrades heavier fractions into lighter, more valuable products
Hydrocracker	It uses hydrogen to upgrade heavier fractions into lighter, more valuable products
Visbreaking	It upgrades heavy residual oils by thermally cracking them into lighter, more valuable reduced viscosity products
Merox	It treats liquefied petroleum gas, kerosene, or jet fuel by oxidizing mercaptans to organic disulfides
Coking	It converts very heavy residual oils into gasoline and diesel fuel, leaving petroleum coke as a residual product
Alkylation	It produces high-octane components for gasoline blending
Dimerization	It converts olefins into higher-octane gasoline blending components
Isomerization	It converts linear molecules to higher-octane branched molecules for blending into gasoline
Steam reforming	It produces hydrogen for the hydrotreaters or hydrocracker
Liquefied gas storage	It is used for propane and similar gaseous fuels at a pressure sufficient to maintain them in liquid form
Storage tanks	They are used for crude oil and finished products, and are usually cylindrical, with some sort of vapor emission control
Amine gas and tail gas treatment	It is used to convert hydrogen sulfide from hydrodesulfurization into elemental sulfur
Utility units	Cooling towers are used to circulate cooling water, and boiler plants are used for steam generation
Wastewater collection and treatment	It converts wastewater to water suitable for reuse or for disposal
Solvent refining	It uses a solvent such as cresol or furfural to remove unwanted, mainly asphaltenic materials from lubricating oil stock
Solvent dewaxing	It removes the heavy waxy constituents of petroleum from vacuum-distillation products

value heavy oil into high-value petrol. For example, catalytic reforming is a conversion process. The purpose of the reformer is to increase the octane number of petrol blend components. Once crude oil has been through the separation and conversion processes, the resulting products are ready for purification, which is principally sulfur removal. A refinery removes sulfur from gasoline, for example, to increase its efficiency and to reduce air pollution from automobile exhausts. Common process units found in an oil refinery are presented in Table 1.3.

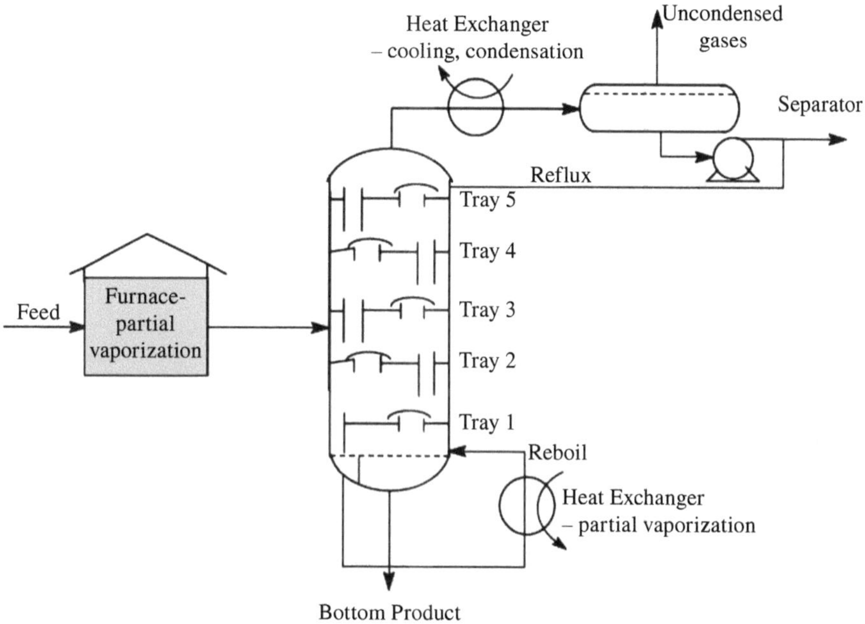

Fig. 1.3 A generalized fractional distillation column

1.2.3 Petroleum Products and Fuels

Petrochemicals are chemical products made from raw materials of petroleum or hydrocarbons of other origin. Although some of the chemical compounds that originate from petroleum may also be derived from other sources such as coal or natural gas, petroleum is a major source of many of them.

The two main classes of petrochemical raw materials are olefins (including ethylene and propylene) and aromatics (including benzene and xylene isomers), both of which are produced in very large quantities. At oil refineries, olefins are produced mainly from hydrocarbons by chemical cracking such as steam cracking and catalytic cracking, whereas aromatic hydrocarbons are mainly produced by catalytic reforming or similar processes. From these basic building blocks, a very wide range of chemicals and other materials used in industry – monomers, solvents, and adhesives – are made. From the monomers, polymers or oligomers are produced for plastics, resins, fibers, elastomers, certain lubricants, and gels.

World production of ethylene is around 110 million tons *per annum*, that of propylene is around 65 million tons *per annum*, and that of aromatic raw materials is about 70 million tons *per annum*. The largest petrochemical industries are to be found in the USA and western Europe, though the major growth in new production capacity is in the Middle East and Asia. There is a substantial interregional trade in petrochemicals of all kinds.

Table 1.4 Classification of oils

	Viscosity (°API)
Conventional oil	
Crude oil	>17
Heavy oil	10–17
Deep-sea oil above 500-m water depth	–
Nonconventional oil	
Extra-heavy oil (including tar sands)	<10
Oil shale	–
Deep-sea oil below 500-m water depth	–
Polar oil product	–
Natural gas liquids	–
Synthetic crude oil	–
Bitumen from tar sands	–

After they have been processed at the refinery, gasoline and other petroleum products are usually shipped out through pipelines. There are about 230,000 miles of pipeline in the USA. Pipelines are the safest and cheapest way to move large quantities of petroleum across land (Demirbas 2009). Pump stations, which are spaced 20–100 miles apart along the underground pipelines, keep the petroleum products moving at around 5 miles per hour. At this rate, it takes 15 days to move a shipment of gasoline from Houston, Texas, to New York City.

Americans use almost 17 million barrels of oil every day of the year. Texas produces more oil than any other state. The other top-producing states are Alaska, California, Louisiana, and Oklahoma, in that order. In all, 31 states produce petroleum.

The USA has become increasingly dependent upon other countries for petroleum. In 1994 the USA purchased 45% of its petroleum from other countries. Americans know this dependence can be dangerous.

The oil industry classifies "crude" by the location of its origin (e.g., West Texas Intermediate or Brent) and often by its relative weight (API gravity) or viscosity ("light," "intermediate," or "heavy"); refiners may also refer to it as "sweet," which means it contains relatively little sulfur, or as "sour," which means it contains substantial amounts of sulfur and requires more refining to meet current product specifications.

There are different classification schemes, based on economic and/or geological criteria. The main oil classes are conventional oil and nonconventional oil. Conventional oil is oil which can be produced with current technology under present economic conditions. Table 1.4 shows the classification of oils.

Gasoline is a petroleum-derived liquid mixture, primarily used as a fuel in internal combustion engines, specifically in spark-ignition engines. In the Otto cycle engine, a mixture of gasoline and air is compressed and is then ignited by a spark plug. The Otto cycle engine was invented by Nikolaus August Otto in 1876; it is also commonly known as a four-stroke cycle engine because of the four strokes involved in an internal combustion engine.

Table 1.5 Physical and chemical properties of gasoline

Property	Information
Color	Colorless to pale brown or pink
Average molecular weight	108
Density (kg/L)	0.7–0.8
Flash point (K)	227.2
Explosive limits in air (%)	1.3–6.0
Flammability limits (%)	1.4–7.4
Autoignition temperature (K)	553–759
Boiling points (K)	
Initially	312
After 10% distillate	333
After 50% distillate	383
After 90% distillate	443
Final boiling point	477
Solubility	
In water at 293 K	Insoluble
In absolute ethanol	Soluble
In diethyl ether	Soluble
In chloroform	Soluble
In benzene	Soluble

Gasoline is a complex mixture of over 500 hydrocarbons that may have between five and 12 carbons. It may also contain some contaminants, including sulfur, nitrogen, oxygen, and certain metals. The four major constituent groups of gasoline are olefins, aromatics, paraffins, and napthenes. It consists mostly of aliphatic hydrocarbons, enhanced with isooctane or the aromatic hydrocarbons toluene and benzene to increase its octane rating.

The density of gasoline is 0.71–0.77 kg/L. Gasoline is more volatile than diesel oil, Jet-A, or kerosene, not only because of the base constituents, but also because of the additives that are put into it. The final control of volatility is often achieved by blending it with butane. The desired volatility depends on the ambient temperature. In hot climates, excessive volatility results in what is known as "vapor lock," where combustion fails to occur, because the liquid fuel has changed to a gaseous fuel in the fuel lines, rendering the fuel pump ineffective and starving the engine of fuel; therefore, gasoline components of higher molecular weight and thus lower volatility are used. In cold climates, too little volatility results in cars failing to start.

The typical composition of gasoline hydrocarbons is as follows: 4–8 vol% alkanes, 2–5 vol% alkenes, 25–40 vol% isoalkanes, 3–7 vol% cycloalkanes, 1–4 vol% cycloalkenes, and 20–50 vol% total aromatics (0.5–2.5 vol% benzene). Additives and blending agents are added to the hydrocarbon mixture to improve the performance and stability of gasoline. These compounds include antiknock agents, antioxidants, metal deactivators, lead scavengers, antirust agents, anti-icing agents, upper-cylinder lubricants, detergents, and dyes. Physical and chemical properties of gasoline are given in Table 1.5. Table 1.6 shows the major components of gasoline.

Table 1.6 Major components of gasoline

Component	Composition by weight (%)
n-Alkanes	
C_5	3.0
C_6	11.6
C_7	1.2
C_9	0.7
C_{10}–C_{13}	0.8
Total of *n*-alkanes	17.3
Branched alkanes	
C_4	2.2
C_5	15.1
C_6	8.0
C_7	1.9
C_8	1.8
C_9	2.1
C_{10}–C_{13}	1.0
Total of branched alkanes	32.0
Cycloalkanes	
C_6	3.0
C_7	1.4
C_8	0.6
Total of cycloalkanes	5.0
Olefins	
C_6	1.8
Total of olefins	
Aromatics	1.8
Benzene	3.2
Toluene	4.8
Xylenes	6.6
Ethylbenzene	1.4
C_3 benzenes	4.2
C_4 benzenes	7.6
Others	2.7
Total aromatics	30.5

Diesel is produced by distilling raw oil which is extracted from bedrock. It is a fossil fuel and consists of hydrocarbons with between nine and 27 carbon atoms in a chain as well as a smaller amount of sulfur, nitrogen, oxygen, and metal compounds. It is a general property of hydrocarbons that the autoignition temperature is higher for more volatile hydrocarbons. The hydrocarbons present in diesel fuel include alkanes, naphthenes, olefins, and aromatics. In addition, other substances are added to improve the characteristics of diesel fuel. Its boiling point is between 445 and 640 K. A good diesel fuel is characterized by low sulfur and aromatic content, good ignition quality, the right cold-weather properties, low content of pollutants, and also the right density, viscosity, and boiling point.

1.2.4 Coal

Both oil and coal have been used in small quantities for thousands of years. But until the Industrial Revolution, society's energy requirements were fulfilled almost entirely by human and animal power and traditional biomass sources. For many years afterward, a large majority of the world remained dependent on traditional biomass. By 1850, population pressures led to the commercialization of coal and the Industrial Revolution, and the energy derived from coal began to shape the forces that would raise the population ceiling. The world's population entered a phase of disequilibrium. Coal also played a large part in the development of electricity. With the establishment of the electricity industry in the 1880s following the remarkable achievements of Edison, Parsons, Stanley, Tesla, Westinghouse, and their collaborators, electricity quickly expanded to power households, industry, and railroads. Electricity was generated in power plants, and those power plants were fed with coal. Still today, 50% of America's power is generated in coal-burning power plants.

Coal can be defined as a sedimentary rock that burns. It was formed by the decomposition of plant matter, and it is a complex substance that can be found in many forms. Coal is divided into four classes: lignite, bituminous coal, subbituminous coal, and anthracite.

Any acceptable scheme of coal classification should be scientific and systematic and should take into account the fundamental characters. The earliest classification of coal recognized the visual characters, agreeing broadly with the properties of combustion. According to this scheme, lignite, or brown coal, is brown to black in color, has a woody structure, and has a high moisture content. Bituminous coal is black, often banded, ignites fairly easily, and burns with flame and smoke. Bituminous coal which has a strong tendency to agglutinate on heating is called caking coal. Anthracite is black and lustrous and difficult to ignite. It burns without a flame because it does not include volatile matter. Elemental analysis gives empirical formulas such as $C_{137}H_{97}O_9NS$ for bituminous coal and $C_{240}H_{90}O_4NS$ for high-grade anthracite.

Anthracite is a dense, hard rock with a jet-black color and a metallic luster. It contains between 86 and 98% carbon by weight, and it burns slowly, with a pale-blue flame and very little smoke. Bituminous coal, or soft coal, contains between 69 and 86% carbon by weight and is the most abundant form of coal. Subbituminous coal contains less carbon and more water, and is therefore a less efficient source of heat. Lignite is a very soft coal that contains up to 70% water by weight. A classification of coal based on the percentage of fixed carbon in the dry, ash-free coal in the coals of higher rank, and the total carbon in those of lower rank is shown in Table 1.7.

Coal is a compact black or dark-brown sedimentary rock that formed some 300 million years ago by high-pressure and high-temperature anaerobic decomposition of dead plants. Plants capture the energy from sunlight through photosynthesis, which directly converts solar energy into plant matter. Animals that then

Table 1.7 Classification of coal based on the percentage of fixed carbon and the total carbon

Class	Fixed carbon (%)	Total carbon (%)
Graphite	>99	–
Anthracite	>93	–
Semianthracite	83–98	–
Semibituminous	73–83	–
Bituminous	47–73	80–88
Subbituminous	47–73	74–80
Cannel	35–47	74–88
Lignite	30–55	65–73
Peat	<50	<65

eat the plants convert that energy again, storing it in their own bodies. Over millions of years, accumulated plant and animal matter is covered by sediment and stored within the earth's crust, gradually being transformed into hard black solids by the sheer weight of the earth's surface.

Coal is the most abundant fossil energy source available to world economies. The use of coal is decreasing in western Europe but is increasing in Asia, where large increases in coal use are projected for the developing countries there (Tillman 2000).

The worldwide coal production is roughly equal to the gas production and only second to that of oil. Coal is produced in deep mines (hard coal) and in surface mines (lignite). Coal has played a key role as a primary source of organic chemicals as well as a primary energy source. Coal may become more important both as an energy source and as the source of carbon-based materials, especially aromatic chemicals in the twenty-first century (Schobert and Song 2002).

Coal accounts for 26% of the world's primary energy consumption and 37% of the energy consumed worldwide for electricity generation. Nearly all the different forms of coal are used in some way or other. For instance, peat has been used for burning in furnaces, lignite is used in power station and home stoves, whereas bituminous coal is used extensively for the generation of electricity.

Despite environmental issues and competitive pressure from other fuels, coal is expected to maintain a major share of the world's future energy use. Large increases in coal use are projected for the developing countries of Asia. Coal's role in energy use worldwide has shifted substantially over the decades, from a fuel used extensively in all sectors of the economy to one that is now used primarily for electricity generation and in a few key industrial sectors, such as steel, cement, and chemicals. Although coal has lost market share to petroleum products and natural gas, it continues to be a key source of energy because of the dominant role it has maintained in its core markets and its success in penetrating markets in emerging economies. For coal to have remained competitive with other sources of energy in the industrialized countries of the world, continuing technological improvements in all aspects of coal extraction have been necessary.

Table 1.8 shows the world's recoverable coal reserves. Coal reserves are rather evenly spread around the globe: 25% are in the USA, 16% are in Russia, and

Table 1.8 World's recoverable coal reserves (million tons) (IEA 2007)

Country	Bituminous including anthracite	Subbituminous	Lignite
USA	115,891	101,021	33,082
China	62,200	33,700	18,600
India	82,396	–	2,000
South Africa	49,520	–	–
Kazakhstan	31,100	–	3,000
Brazil	–	11,929	–
Colombia	6,267	381	–
Canada	3,471	871	2,236
Indonesia	790	1,430	3,150
Botswana	4,300	–	–
Uzbekistan	1,000	–	3,000
Turkey	278	761	2,650
Pakistan	–	2,265	–
Thailand	–	–	1,268
Chile	31	1,150	–
Mexico	860	300	51
Peru	960	–	100
Kyrgyzstan	–	–	812
Japan	773	–	–
North Korea	300	300	–
Zimbabwe	502	–	–
Venezuela	479	–	–
The Philippines	–	232	100
Mozambique	212	–	–
Swaziland	208	–	–
Tanzania	200	–	–
Others	449	379	27

11.5% are in China. Although coal is much more abundant than oil and gas on a global scale, coalfields can be depleted on a regional scale.

1.2.5 Chemicals and Fuels from Coal

As early as 1800, coal gas, or town gas, was made by heating coal in the absence of air. An ideal coal gasification process would convert the coal completely to gas. Coal can be converted to water gas with steam (Demirbas 2007). The key chemical step for achieving this is the reaction

$$C + H_2O \rightarrow CO + H_2, \qquad (1.1)$$

which takes place at temperatures above 1,075 K. Water gas burns to give CO_2 and H_2O, releasing roughly 11.2 kJ per liter of gas consumed.

Water gas formed by the reaction of coal with oxygen and steam is a mixture of CO, CO_2, and H_2. The ratio of H_2 to CO can be increased by adding water to this mixture, to take advantage of a reaction known as the water-gas shift reaction (Demirbas 2007).

$$CO + H_2O \rightarrow CO_2 + H_2 \quad (1.2)$$

The concentration of CO_2 can be decreased by reacting the CO_2 with coal at high temperatures to form CO.

$$C + CO_2 \rightarrow 2CO \quad (1.3)$$

Water gas from which the CO_2 has been removed is called synthesis gas (syngas) because it can be used as a starting material for a variety of organic and inorganic compounds. It can be used as the source of H_2 for the synthesis of methanol, for example.

$$CO + 2H_2 \rightarrow CH_3OH \quad (1.4)$$

Methanol can then be used as a starting material for the synthesis of alkenes, aromatic compounds, acetic acid, formaldehyde, and ethyl alcohol (ethanol). Syngas can also be used to produce methane, or synthetic natural gas (Demirbas 2007).

The CO/H_2 mixture cannot be used as a direct replacement for methane. An additional process, methanation, is needed:

$$CO + 3H_2 \rightarrow CH_4 + H_2O, \quad (1.5)$$

$$2CO + 2H_2 \rightarrow CH_4 + CO_2 \quad (1.6)$$

The methane produced in this way is sometimes called substitute natural gas or synthetic natural gas. It would be easier, and possibly cheaper, to obtain synthetic natural gas in a single processing step:

$$C + 2H_2 \rightarrow CH_4. \quad (1.7)$$

This process concept is called hydrogasification. It has not been used commercially, but has been the subject of much research and development effort. A promising new development in coal gasification is the integrated gasification combined cycle plant. It is being advertised by the electric power industry as the "cleanest method available for making electricity from coal."

The first step toward making liquid fuels from coal involves the manufacture of syngas (CO and H_2) from coal. Coal can be gasified to produce syngas, which can be converted to paraffinic liquid fuels and chemicals by Fischer–Tropsch synthesis, which was developed in 1925 by Franz Fischer and Hans Tropsch to converted CO and H_2 at 1 atm and 575 K into liquid hydrocarbons using an Fe/Co catalyst. By 1941, Fischer–Tropsch plants produced 740,000 tons of petroleum products per year in Germany (Dry 1999). Fischer–Tropsch technology is based on a complex series of reactions that use H_2 to reduce CO to CH_2 groups linked to form long-chain hydrocarbons (Schulz 1999).

$$nCO + (n + m/2)\,H_2 \rightarrow C_nH_m + nH_2O \quad (1.8)$$

At the end of World War II, Fischer–Tropsch technology was being studied in most industrial nations. The liquid product mainly contains benzene, toluene, xylene, phenols, alkylphenols, and cresol. The low cost and high availability of crude oil, however, led to a decline in interest in liquid fuels made from coal.

1.2.6 Natural Gas

The role of natural gas in the world's energy supply is growing rapidly. Growing demand for natural gas is expected in all regions of the world, as resource availability, cost, and environmental considerations all favor its use.

The reserves and resources of conventional natural gas are comparable in size to those of conventional oil, but global gas consumption is still considerably lower than that of oil. The proved gas reserves are not evenly distributed around the globe: 41% of them are in the Middle East and 27% of them are in Russia. A peak in conventional gas production may occur between 2020 and 2050. Natural gas accounts today for 25% of world primary energy production. Natural gas is the fastest-growing primary energy source in the world. Because it is a cleaner fuel than oil or coal and not as controversial as nuclear power, gas is expected to be the fuel of choice for many countries in the future. The progress achieved in reducing the cost and increasing the performance of gas turbines, the present low prices, and the large world reserves make natural gas a highly desired fuel.

1.2.7 Gas Hydrates

Methane can be obtained from gas hydrates. Gas hydrates are also called methane hydrates or chemically clathrates. Gas hydrates are potentially one of the most important energy resources for the future. Methane gas hydrates are increasingly considered to be a potential energy resource. Methane gas hydrates are crystalline solids formed by the combination of methane and water at low temperatures and high pressures. Gas hydrates have an icelike crystalline lattice of water molecules with methane molecules trapped inside. Enormous reserves of hydrates can be found under continental shelves and on land under permafrost. The amount of organic carbon in gas hydrates is estimated to be twice that in all other fossil fuels combined. However, owing to the solid form of the gas hydrates, conventional gas and oil recovery techniques are not suitable. Table 1.9 shows worldwide amounts of organic carbon sources. The recovery of methane generally involves dissociating or melting *in situ* gas hydrates by heating the reservoir above the temperature of hydrate formation, or decreasing the reservoir pressure below that of hydrate equilibrium (Lee and Holder 2001).

Table 1.9 Worldwide amounts of organic carbon sources (Hacisalihoglu *et al.* 2008)

Source of organic carbon	Amount (gigatons)
Gas hydrates (onshore and offshore)	10,000–11,000
Recoverable and nonrecoverable fossil fuels (oil, coal, gas)	5,000
Soil	1,400
Dissolved organic matter	980
Land biota	880
Peat	500
Others	70

1.2.8 Oil Shale

Oil shale, or shale oil, is a inorganic, nonporous sedimentary rock that contains the solid hydrocarbon wax kerogen in tightly packed limy mud and clay. In some respects, oil shale is similar to the source rock that produced petroleum. Oil shale contains more kerogen than oil source rock, but less than coal. The composition of the oil derived from oil shale is much more similar to the composition of petroleum than to that of coal. Some oil shales can be ignited, like coal, and they burn with a very sooty, smoky flame, much like a coal of very high volatile matter content. However, oil shales are of no interest as solid fuels. The principal interest in them is in the possibility of their conversion to liquid fuels.

Two thirds of the world's oil shale reserves are located in the USA. The largest known reserves of hydrocarbons of any kind are the Green River shale deposits in Wyoming, Colorado, and Utah. These reserves are estimated to be 270 billion tons. At 20 gallons per ton of shale, this translates into 130 billion barrels of oil. This is 5 times as much as the proved reserves of petroleum in the USA. However, no commercial production of fuels from oil shale exists today, so their economic recoverability is not well known.

The kerogen may be decomposed at elevated temperatures (723 K), resulting in an oil suitable for refinery processing (Dorf 1977). The oil shale layer was not hot enough to complete oil generation. For the final step, the kerogene must be heated to 775 K and molecularly combine with additional hydrogen to complete the oil formation. This final process must be performed in the refinery and needs huge amounts of energy, which otherwise were provided by the geological environment during oil formation (Demirbas 2000). The kerogene is still in the source rock and could not accumulate in oil fields. Typically, the ratio of kerogene to waste material is very low, making the mining of oil shales unattractive. Hence, owing to a combination of environmental and economic concerns, it is very unlikely that oil shale mining will ever be performed on a large scale, though in some places it has been utilized in small quantities. However, the world's shale oil reserves are greater than those of crude oil or natural gas, as shown in Fig. 1.1.

The industrial processing of oil shale began a century ago. During the 1920s, oil shale was an economically important energy source. During the energy crisis of the 1970s, several major oil companies made massive investments both in research and development and in possible commercial use of oil shale. Several technical problems must be overcome for successful large-scale use of oil shale in the future. The composition of oil from oil shale is sufficiently different from that of petroleum that liquids derived from oil shale cannot be used as direct substitutes for petroleum. Oil derived from oil shale has less carbon and hydrogen and more nitrogen and oxygen than petroleum. Petroleum refinery operations would have to be modified to accommodate oil shale liquids as a feedstock. Hydrogen must be added to the oil during processing. When oil shale is retorted, the inorganic portion of the shale expands considerably. The liquid fuels derived from oil shale could become important alternatives to petroleum.

1.2.9 Tar Sands

Tar sands are grains of sand or, in some cases, porous carbonate rocks that are intimately mixed with a very heavy, asphalt-like crude oil called bitumen. The bitumen is much too viscous to be recovered by traditional petroleum recovery techniques. Tar sands contain about 10–15% bitumen, the remainder being sand or other inorganic materials.

The estimated worldwide resources of tar sands are about 3 times the known petroleum reserves. The world's largest deposit of tar sands is near Athabasca, in Alberta, Canada. Other large deposits exist in the former Soviet Union and in Venezuela. In the USA, small deposits of tar sands are found in Utah.

Tar sands are oil traps not deep enough in the earth to allow for geological conversion into conventional oil. This oil was not heated enough to complete the process of molecular breakage to reduce the viscosity. The oil has the characteristics of bitumen and is mixed with large amounts of sand owing to the proximity to the earth's surface. The tar sand is mined, flooded with water to separate the heavier sand, and then processed in special refineries to reduce its high sulfur content (the original oil usually has 3–5% sulfur) and other components. This process needs huge amounts of energy and water. Only oil deposits in deep layers below 75 m are mined *in situ* (COSO 2007).

If tar sand is heated to about 355 K, by injecting steam into the deposit in a manner analogous to that of enhanced oil recovery, the elevated temperature causes a decrease in the viscosity of the bitumen just enough to allow its pumping to the surface. Alternatively, it is sometimes easier to mine the tar sand as a solid material. If the bitumen is heated above 775 K, about 70% of it is converted to a synthetic crude oil. Distilling this oil gives good yields of kerosene and other liquid products in the middle distillate range.

1.2.10 Natural Bitumen and Extra-heavy Oil

Natural bitumen is the oil contained in clastic and carbonate reservoir rocks, most frequently in small deposits at, or near, the earth's surface. Clastic rocks are composed of fragments of preexisting rock. These rocks are commonly referred to as tar sands or oil sands and have been mined since antiquity for use as paving. Natural bitumen is found in 183 identified deposits in 21 countries. Natural bitumen and heavy oil differ from light oils in their high viscosity at reservoir temperatures, high density (low API gravity), and significant contents of nitrogen, oxygen, and sulfur compounds and heavy-metal contaminants.

Historically, heavy oil was found incidentally during the search for light oil and was produced by conventional methods when economically feasible. However, to sustain commercial well production rates, heavy and extra-heavy oil production almost always requires measures to reduce oil viscosity and to introduce energy into the reservoir. Extra-heavy oil commonly requires the addition of diluents (gas condensate, natural gas liquids, or light crude) to enable the oil to be transported by pipeline. Extra-heavy oil must also be chemically upgraded to reduce its density and remove contaminants before it can be used as a refinery feedstock.

Extra-heavy oil and natural bitumen represent crude oils which have been severely degraded by microbial action, as evidenced by their paucity of low molecular weight saturated hydrocarbons. Extra-heavy oil is recorded in 91 deposits. Some of these represent separate reservoirs in a single field, of which some are producing and others have been abandoned.

The recoverable volumes of oil contained in deposits of extra-heavy oil and natural bitumen are immense. Figure 1.4 provides a comparison of the distribution of the world's known recoverable oil resources and reserves by type. Table 1.10 shows comparative chemical analyses of world oils and natural bitumen.

Two fundamental upgrading processes are presently employed to prepare heavy oil and natural bitumen for transportation and refining to finished products. These

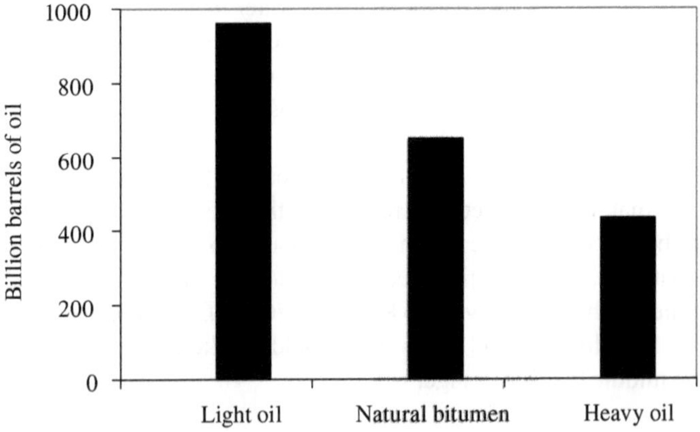

Fig. 1.4 Distribution of the world's known recoverable oil resources and reserves by type

Table 1.10 Comparative chemical analyses of world oils and natural bitumen

	Conventional oil	Medium oil	Heavy oil	Extra-heavy oil	Natural bitumen
Coke in crude oil (wt%)	10.4	17.6	21.8	28.1	–
Gasoline yields (vol%)	9.2	2.8	2.0	1.3	1.4
Gas oil yield (vol%)	17.4	21.9	15.9	16.9	7.2
Residuum yield (vol%)	21.9	39.5	52.6	62.6	18.1
Crude oil density (kg/L)	0.836	0.920	0.958	1.018	1.041
Crude oil gravity (°API)	38.1	22.3	16.3	7.5	5.0
Crude oil viscosity (cP at 411 K)	9	63	593	7,936	292,991
Resins (wt%)	6.1	19.3	24.2	21.2	25.2
Asphaltenes (wt%)	2.1	6.6	12.4	13.2	30.6
Sulfur (wt%)	0.4	1.5	2.9	4.9	3.3

processes are carbon rejection and hydrogen addition. Each process improves the hydrogen-to-carbon ratio, but by following different paths. Carbon rejection yields a large quantity of low-heating-value gas at the expense of produced liquids, a large amount of petroleum coke, and therefore moderate conversion at low pressure. The yield of upgraded oil (synthetic crude oil) from the natural bitumen varies with the technology employed, the consumption of product for the fuel in the upgrader, the extent of natural gas liquid recovery, and the degree of residue upgrading.

The densities and viscosities are directly related to their origin as the degraded residues of conventional oils. This has resulted in a large increase in the content of polar molecules, of which the asphaltenes are the largest and most troublesome.

Natural bitumen is so viscous that it is immobile in the reservoir. For oil sand deposits less than 225 feet deep, bitumen is recovered by mining the sands, then separating the bitumen from the reservoir rock by processing it with hot waters, and finally upgrading the natural bitumen on-site to a synthetic crude oil. In deeper oil sand deposits, where the bitumen is commonly less viscous, steam is injected into the reservoir to mobilize the oil for recovery by production wells. The product may be upgraded on-site or mixed with diluent and transported to an upgrading facility.

1.3 Renewable Energy Sources

The term "renewable energy" refers to primary energies that are continuously generated by solar, geothermal, and tidal processes. The energy produced within the sun is responsible for a multitude of other renewable energies (such as wind and hydropower) as well as renewable energy carriers (such as solid or liquid biofuels). The main renewable energy sources are biomass, hydropower, geothermal, solar, wind, and marine energies. The renewables are the primary, domestic and clean or inexhaustible energy resources.

Renewable energy technologies produce marketable energy by converting natural phenomena into useful forms of energy. These technologies use the sun's energy and its direct and indirect effects on the earth (solar radiation, wind, falling water, and various plants, i.e., biomass), gravitational forces (tides), and the heat of the earth's core (geothermal) as the resources from which energy is produced. The three types of renewable energy give rise to a multitude of very different energy flows and carriers owing to various energy conversion processes occurring in nature.

Renewable energy is a promising alternative solution because it is clean and environmentally safe. Renewable energy sources also produce lower or negligible levels of greenhouse gases and other pollutants when compared with the fossil energy sources they are replacing.

Biomass is a combustible renewable energy source. Hydropower, geothermal, wind, solar, wave, tidal, and ocean thermal energy are noncombustible renewable energy sources. Noncombustible renewable energies have been the primary energy source in the history of the human race.

1.3.1 Biomass

The term "biomass" (Greek *bios*, meaning "life," and "*maza*, meaning "mass") refers to nonfossilized and biodegradable organic material originating from plants, animals, and microorganisms derived from biological sources. The biomass includes products, by-products, residues, and waste from agriculture, forestry, and related industries as well as the nonfossilized and biodegradable organic fractions of industrial and municipal solid waste. The term "biomass" refers to wood, short-rotation woody crops, agricultural waste, short-rotation herbaceous species, wood waste, bagasse, industrial residues, waste paper, municipal solid waste, sawdust, biosolids, grass, waste from food processing, aquatic plants, and algal animal waste, and a host of other materials.

Biomass is organic material that has stored sunlight in the form of chemical energy. Biomass is commonly recognized as an important renewable energy source; during the growth of plants and trees, solar energy is stored as chemical energy via photosynthesis, which can be released via direct or indirect combustion.

Biomass is very important for implementing the Kyoto agreement to reduce carbon dioxide emissions by replacing fossil fuels. Rapidly increasing energy requirements parallel technological developments, and research and development activities are forced to study new and biorenewable energy sources.

There are three ways to use biomass. It can be burned to produce heat and electricity, changed to gaslike fuels such as methane, hydrogen, and carbon monoxide, or changed to a liquid fuel. Liquid fuels, also called biofuels, include mainly two forms of alcohol: ethanol and methanol. Because biomass can be changed directly into a liquid fuel, it could someday supply much of our transportation fuel needs for cars, trucks, buses, airplanes, and trains. This is very important because nearly one third of the USA's energy is now used for transportation.

Table 1.11 Major categories of biomass feedstocks

Forest products	Wood, logging residues, trees, shrubs, wood residues, sawdust, bark, etc.
Biorenewable waste	Agricultural waste, crop residues, mill wood waste, urban wood waste, urban organic waste
Energy crops	Short-rotation woody crops, herbaceous woody crops, grasses, starch crops, sugar crops, forage crops, oilseed crops, switchgrass, *Miscanthus*
Aquatic plants	Algae, water weed, water hyacinth, reeds, and rushes
Food crops	Grains, oil crops
Sugar crops	Sugarcane, sugar beets, molasses, sorghum
Landfill	Hazardous waste, nonhazardous waste, inert waste, liquid waste
Organic waste	Municipal solid waste, industrial organic waste, municipal sewage and sludges
Algae	Prokaryotic algae, eukaryotic algae, kelps
Mosses	Bryophyta, Polytrichales
Lichens	Crustose lichens, foliose lichens, fruticose lichen

Biomass feedstocks are marked by their tremendous diversity, which makes them rather difficult to characterize as a whole. Feedstocks that can be utilized with conversion processes are primarily the organic materials now being disposed of in landfills. These include forest product waste, agricultural residues, organic fractions of municipal solid waste, paper, cardboard, plastic, food waste, green waste, and other waste. Nonbiodegradable organic feedstocks, such as most plastics, are not convertible by biochemical processes. Bio-based materials require pretreatment by chemical, physical, or biological means to open up the structure of biomass. The general categories of biomass feedstock are shown in Table 1.11.

The importance of biomass in different world regions is given in Table 1.12. For large portions of the rural populations of developing countries, and for the poorest sections of urban populations, biomass is often the only available and affordable source of energy for basic needs such as cooking and heating. As shown in Table 1.12, the importance of biomass varies significantly across regions. In Europe, North America, and the Middle East, the share of biomass averages 2–3% of total final energy consumption, whereas in Africa, Asia, and Latin America, which together account for three quarters of the world's population, biomass provides a substantial share of the energy needs: one third on average, but as much as 80–90% in some of the poorest countries of Africa and Asia (e.g., Angola, Ethiopia, Mozambique, Tanzania, Democratic Republic of Congo, Nepal, and Myanmar).

The basic structure of lignocellulosic biomass, such as all wood and woody plants, agricultural residues, and crops, consists of cellulose, hemicelluloses, lignin, extractive lipids, proteins, simple sugars, starches, water, hydrocarbons, ash, and other compounds. Two larger carbohydrate categories that have significant value are cellulose and hemicelluloses (holocellulose). The lignin fraction consists of non-sugar-type macromolecules. Three structural components are cellulose, hemicelluloses, and lignin, which roughly have the formulas $CH_{1.67}O_{0.83}$, $CH_{1.64}O_{0.78}$, and $C_{10}H_{11}O_{3.5}$, respectively.

Table 1.12 The importance of biomass in different world regions

Region	Share of biomass in final energy consumption (%)
Africa (average)	62.0
Burundi	93.8
Ethiopia	85.6
Kenya	69.6
Somalia	86.5
Sudan	83.7
Uganda	94.6
South Asia (average)	56.3
East Asia (average)	25.1
China	23.5
Latin America (average)	18.2
Europe (average)	3.5
North America (average)	2.7
Middle East (average)	0.3

Cellulose is a linear polymer composed of repeating anhydroglucose units. It is a remarkable pure organic polymer, consisting solely of units of anhydroglucose held together in a giant straight-chain molecule (Demirbas 2000). These anhydroglucose units are bound together by β-(1,4)-glycosidic linkages. Owing to this linkage, cellobiose is established as the repeat unit for cellulose chains. Cellulose must be hydrolyzed to glucose before fermentation to ethanol.

The second major chemical species in wood are the hemicelluloses. They are amorphous polysaccharides, such as xylans, galactoglucomannans, arabinogalactans, glucans, and galactans. The hemicelluloses, unlike cellulose, not only contain glucose units, but they are also composed of a number of different pentose and hexose monosaccharides. Hemicelluloses tend to be much shorter than cellulose, and the molecular structure is slightly branched.

Unlike cellulose, hemicelluloses consist of different monosaccharide units. In addition, the polymer chains of hemicelluloses have short branches and are amorphous. Because of the amorphous morphology, hemicelluloses are partially soluble or swellable in water. Hemicelluloses (arabinoglycuronoxylan and galactoglucomammans) are related to plant gums in composition, and occur in much shorter molecular chains than cellulose. Hemicelluloses are derived mainly from chains of pentose sugars, and act as the cement material holding together the cellulose micelles and fiber (Theander 1985).

Lignin is a chemical compound that is most commonly derived from wood and is an integral part of the cell walls of plants, especially in tracheids, xylem fibers, and sclereids. It is one of most abundant organic compounds on the earth after cellulose and chitin. Lignin is a complex, high molecular weight polymer built of hydroxyphenylpropane units. It is a large, cross-linked macromolecule which consists of alkylphenols and has a complex three-dimensional structure. Its functions are to provide structural strength, provide sealing of the water-conducting system that links roots with leaves, and protect plants against degradation (Glasser and Sarkanen 1989). It is lignin that holds wood cells together and provides the extraordinary composite strength characteristics of a piece of wood.

Lignin has a molecular mass in excess of 10,000 amu. The molecular mass in nature is difficult to measure, since lignin is fragmented during preparation. It is relatively hydrophobic and aromatic in nature. The molecule consists of various types of substructures which appear to repeat in a random manner. Lignin plays a crucial part in conducting water in plant stems. The polysaccharide components of plant cell walls are highly hydrophilic and thus permeable to water, whereas lignin is more hydrophobic. The cross-linking of polysaccharides by lignin is an obstacle for absorption of water to the cell wall. Thus, lignin makes it possible for plants to form a vascular network which conducts water efficiently. Lignin is present in all vascular plants, but not in bryophytes, supporting the idea that the main function of lignin is related to water transport. Lignin is covalently linked with xylans in the case of hardwoods and with galactoglucomannans in softwoods (Sarkanen and Ludwig 1971).

Pyrolysis of lignin, e.g., during combustion, yields a range of products, of which the most characteristic ones are methoxy phenols. Of those, the most important are guaiacol and syringol and their derivatives; their presence can be used to trace a smoke source to a wood fire. Lignin biosynthesis begins with the synthesis of monolignols. The starting material is the amino acid phenylalanine. The first reactions in the biosynthesis are shared with the phenylpropanoid pathway, and monolignols are considered to be part of this group of compounds. There are three main types of monolignols: coniferyl alcohol, sinapyl alcohol, and paracoumaryl alcohol. Different plants use different monolignols. For example, gymnosperms such as Norway spruce have a lignin that consists almost entirely of coniferyl alcohol. Dicotyledonous lignin is a mixture of coniferyl alcohol and sinapyl alcohol (normally more of the latter), and monocotylednous lignin is a mixture of all three monolignols. Some monocotyledons, such as many grasses, have mostly coniferyl alcohol, whereas others, such as some palms, have mainly sinapyl alcohols. Monolignols are synthesized in the cytosol as glucosides. The glucose is added to the monolignols to make them water-soluble and to reduce their toxicity. The glucose is then removed and the monolignols are polymerized into lignin. Four of the main monolignols in the lignin structure are given in Fig. 1.5.

Wood and woody biomass also contain lesser amounts of tannins, simple sugars, starches, pectins, and organic soluble extractives. Extractives include terpenes, tall oil, and the fatty acids, esters, and triglycerides which contribute to paper mill pitch problems (Demirbas 1991).

In general, combustion models of biomass can be classified as macroscopic or microscopic. The characterization of the biomass fuels is generally divided into three broad categories: (1) composition and structure; (2) reactivity measures; and (3) ash chemistry.

The fuel characteristics of biomass, such as ultimate analysis, heating value, moisture content, particle size, bulk density, and ash fusion temperature, have been reviewed (Bushnell *et al.* 1989). Fuel properties for the combustion analysis of biomass can be conveniently grouped into physical, chemical, thermal, and mineral properties (Ragland *et al.* 1991).

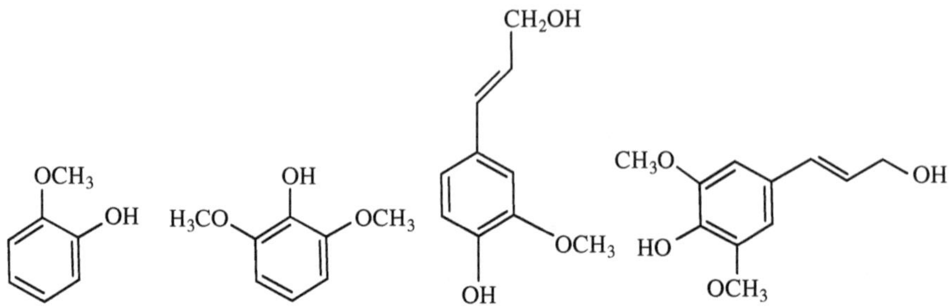

Guaiacol Syringol Coniferyl alcohol Sinapyl alcohol

Fig. 1.5 Four of the main monolignols in the lignin structure

Physical property values vary greatly and properties such as density, porosity, and internal surface area depend on the biomass species, whereas bulk density, particle size, and shape distribution depend on the fuel preparation methods. Important chemical properties for combustion are the ultimate analysis, proximate analysis, analysis of pyrolysis products, higher heating value, heat of pyrolysis, heating value of the volatiles, and heating value of the char.

Thermal property values such as specific heat, thermal conductivity, and emissivity vary with moisture content, temperature, and the degree of thermal degradation by 1 order of magnitude. Thermal degradation products of biomass consist of moisture, volatiles, char, and ash. Volatiles are further subdivided into gases such as light hydrocarbons, carbon monoxide, carbon dioxide, hydrogen, and moisture, and tars. The yields depend on the temperature and heating rate of pyrolysis. Some properties vary with species, location within the biomass, and growth conditions. Other properties depend on the combustion environment. Where the properties are highly variable, the likely range of the property is given (Ragland *et al.* 1991).

Biomass can be converted into useful biofuels and biochemicals via biomass upgrading and biorefinery technologies. Biomass conversion processes include fractionation, densification (briquetting, pelleting), liquefaction, supercritical fluid liquefaction, destructive carbonization, pyrolysis, gasification, hydrothermal liquefaction and hydrothermal upgrading, Fischer–Tropsch synthesis, anaerobic digestion, hydrolysis, and fermentation. Figure 1.6 shows the main biomass conversion processes.

Direct combustion is the old way of using biomass. Biomass thermochemical conversion technologies such as pyrolysis and gasification are certainly not the most important options at present; combustion is responsible for over 97% of the world's bioenergy production (Demirbas 2004). Direct combustion and co-firing with coal for production of electricity from biomass has been found to be a promising method. The energy used for cooking and heating is dominated by energy produced from traditional biomass, especially in rural areas of developing countries. Traditional biomass cooking and heating produces high levels of pollutants.

The main transportation fuels that can be obtained from biomass using different processes are sugar ethanol, cellulosic ethanol, grain ethanol, biodiesel, pyrolysis

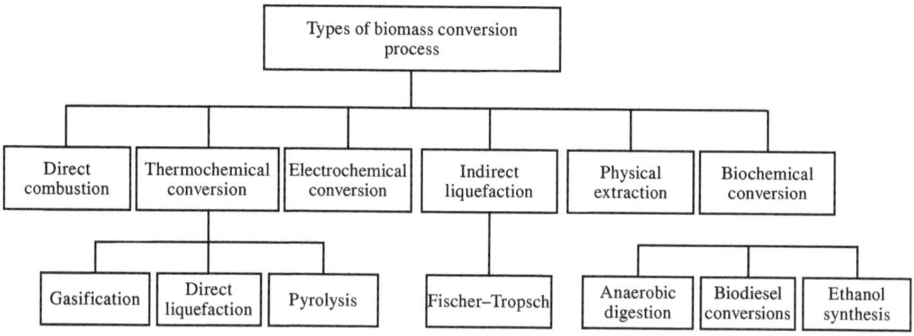

Fig. 1.6 Main biomass conversion processes

liquids, green diesel, green gasoline, butanol, methanol, syngas liquids, biohydrogen, algae diesel, algae jet fuel, and hydrocarbons.

There are several ways to make use of the energy contained in the biomass, from old direct burning to pyrolysis, gasification, and liquefaction.

In the liquefaction process, biomass is converted to liquefied products through a complex sequence of physical structure and chemical changes. The feedstock of liquefaction is usually a wet matter. In the liquefaction, biomass is decomposed into small molecules. These small molecules are unstable and reactive, and can repolymerize into oily compounds with a wide range of molecular weight distribution (Demirbas 2000).

Pyrolysis is the basic thermochemical process for converting biomass to a more useful fuel. Biomass is heated in the absence of oxygen, or is partially combusted in a limited oxygen supply, to produce a hydrocarbon-rich gas mixture, an oil-like liquid, and a carbon-rich solid residue. Rapid heating and rapid quenching produce the intermediate pyrolysis liquid products, which condense before further reactions break down higher molecular weight species into gaseous products. High reaction rates minimize char formation. Under some conditions, no char is formed. At higher fast pyrolysis temperatures, the major product is gas.

Pyrolysis is the simplest and almost certainly the oldest method of processing one fuel to produce a better one. Pyrolysis can also be carried out in the presence of a small quantity of oxygen (gasification), water (steam gasification), or hydrogen (hydrogenation). One of the most useful products is methane, which is a suitable fuel for electricity generation using high-efficiency gas turbines.

Cellulose and hemicelluloses form mainly volatile products on heating owing to the thermal cleavage of the sugar units. The lignin forms mainly char since it is not readily cleaved to lower molecular weight fragments. The progressive increase in the pyrolysis temperature of the wood leads to the release of the volatiles, thus forming a solid residue that is different chemically from the original starting material (Demirbas 2000). Cellulose and hemicelluloses initially break into compounds of lower molecular weight. This forms an "activated cellulose" which decomposes by two competitive reactions: one forming volatiles (anhydrosugars) and the other forming char and gases. The thermal degradation of the activated cellulose and hemicelluloses to form volatiles and char can be divided into categories depending on the reaction temperature. Within a fire, all these reactions take place concurrently and consecutively. Gaseous emissions are predominantly a product of pyrolytic cracking of the fuel. If flames are present, fire temperatures are high, and more oxygen is available from thermally induced convection.

The biomass pyrolysis is attractive because solid biomass and waste can be readily converted into liquid products. These liquids, such as crude bio-oil or slurry of charcoal of water or oil, have advantages in transport, storage, combustion, retrofitting, and flexibility in production and marketing.

Gasification is a form of pyrolysis carried out in the presence of a small quantity of oxygen at high temperatures to optimize the gas production. The resulting gas, known as producer gas, is a mixture of carbon monoxide, hydrogen, and methane, together with carbon dioxide and nitrogen. The gas is more versatile than the original solid biomass (usually wood or charcoal): it can be burned to produce process heat and steam, or used in gas turbines to produce electricity.

Biomass gasification technologies are expected to be an important part of the effort to meet these goals of expanding the use of biomass. Gasification technologies provide the opportunity to convert renewable biomass feedstocks into clean fuel gases or syngases. Biomass gasification is the latest generation of biomass energy conversion processes, and is being used to improve the efficiency and to reduce the investment costs of biomass electricity generation through the use gas turbine technology. High efficiencies (up to about 50%) are achievable using combined-cycle gas turbine systems, where waste gases from the gas turbine are recovered to produce steam for use in a steam turbine.

Various gasification technologies include gasifiers where the biomass is introduced at the top of the reactor and the gasifying medium is either directed in the same direction (downdraft) or in the opposite direction up through the packed bed (updraft). Other gasifier designs incorporate circulating or bubbling fluidized beds. Tar yields can range from 0.1% (downdraft) to 20% (updraft) or greater in the product gases.

The process of obtaining synthetic fuels (synfuels) from biomass will lower the energy cost, improve the waste management, and reduce harmful emissions. This triple assault on plant operating challenges is a proprietary technology that gasifies biomass by reacting it with steam at high temperatures to form a clean burning synthetic gas (syngas, $CO + H_2$). The molecules in the biomass (primarily carbon, hydrogen, and oxygen) and the molecules in the steam (hydrogen and oxygen) reorganize to form this syngas.

1.3.1.1 Marine Biomass

New energy crops, including marine biomass, will need to be introduced. Marine biomass has attracted less attention than terrestrial biomass for energy utilization so far, but is worth considering especially for a country such as Japan which has long available coastlines.

Plankton, algae, and other marine-based organisms constitute a biomass resource that has not yet been exploited. This area is, however, the subject of continued research. Bearing in mind the volume of the sea, this resource could provide a major carbon-neutral source of energy for the future. The productive ecosystems in the world total an estimated 50 billion hectares, excluding the ice caps. Marine ecosystems occupy approximately 36.5 billion hectares, whereas the terrestrial ecosystems occupy approximately 13.5 billion hectares.

Algae are simple organisms that are mainly aquatic and microscopic. Microalgae are unicellular photosynthetic microorganisms living in saline or freshwater environments that convert sunlight, water, and carbon dioxide to algal biomass. They are categorized into four main classes: diatoms, green algae, blue-green algae, and golden algae. There are two main populations of algae: filamentous and phytoplankton algae. These two species, in particular phytoplankton, increase in numbers rapidly to form algal blooms. Seawater supplemented with commercial nitrate and phosphate fertilizers and a few other micronutrients is commonly used for growing marine microalgae (Molina Grima *et al.* 1999).

Macroalgae, more commonly known as "seaweed," are fast-growing marine and freshwater plants that can grow to considerable size (up to 60 m in length). Emergents are plants that grow partially submerged in bogs and marshes. Microalgae are, as the name suggests, microscopic photosynthetic organisms. Like macroalgae, these organisms are found in both marine and freshwater environments.

1.3.1.2 Biofuels from Biomass

The term "biofuel" refers to solid, liquid, or gaseous fuels that are predominantly produced from biorenewable or combustible renewable feedstocks. Liquid biofuels will be important in the future because they will replace petroleum fuels. The biggest difference between biofuels and petroleum feedstocks is the oxygen content. Biofuels are nonpolluting, locally available, accessible, sustainable, and reliable fuels obtained from renewable sources. Biofuels can be classified on the basis of their production technologies: first-generation biofuels, second-generation biofuels, third-generation biofuels, and fourth-generation biofuels.

The first-generation biofuels refer to biofuels made from sugar, starch, vegetable oils, or animal fats using conventional technology. The basic feedstocks for the production of first-generation biofuels are often seeds or grains such as wheat, which yields starch that is fermented into bioethanol, or sunflower seeds, which are pressed to yield vegetable oil that can be used in biodiesel.

Second- and third-generation biofuels are also called advanced biofuels. Second-generation biofuels are made from nonfood crops, wheat straw, corn, and wood using advanced technology. Algae fuel, also called algal oil or a third-generation biofuel, is a biofuel from algae. Algae are low-input/high-yield (30 times more energy per acre than land) feedstocks to produce biofuels using more advanced technology. On the other hand, an appearing fourth generation is based in the conversion of vegoil and biodiesel into biogasoline using the most advanced technology.

There are some barriers to the development of biofuel production. They are technological, economic, supply, storage, safety, and policy barriers. Reducing these barriers is one of the driving factors in the US government's involvement in biofuel and biofuel research and development. Production costs are uncertain and vary with the feedstock available. The production of biofuels from lignocellulosic feedstocks can be achieved through two very different processing routes: biochemical and thermochemical. There is no clear candidate for the "best technology pathway" between the competing biochemical and thermochemical routes. Technical barriers for enzymatic hydrolysis include low specific activity of current commercial enzymes, high cost of enzyme production, and lack of understanding of enzyme biochemistry and mechanistic fundamentals.

The major nontechnical barriers are restrictions or prior claims on use of land (food, energy, amenity use, housing, commerce, industry, leisure, or designations as areas of natural beauty, special scientific interest, etc.), as well as the environmental and ecological effects of large areas of monoculture. For example, vegetable oils are a renewable and potentially inexhaustible source of energy with an energy content close to that of diesel fuel. On the other hand, extensive use of vegetable oils may cause other significant problems, such as starvation in developing countries. The vegetable oil fuels were not acceptable because they were more expensive than petroleum fuels.

There are few technical barriers to building biomass-fired facilities at any scale, from domestic scale to around 50 MW, above which considerations of the availability and cost of providing fuel become significant. In general, however, the capacity and generating efficiency of biomass plants are considerably less than those of modern natural-gas-fired turbine systems. The main nontechnical limitations to investment in larger systems are economic, or in some countries reflect planning conditions and public opinion, where a clear distinction may not be made between a modern effective biomass energy plant and older polluting incinerator designs.

There are two global biorenewable liquid transportation fuels that might replace gasoline and diesel fuel. These are bioethanol and biodiesel. Bioethanol is good alternative fuel that is produced almost entirely from food crops. Biodiesel has become more attractive recently because of its environmental benefits.

Transport is one of the main energy-consuming sectors. It is assumed that biodiesel will be used as a fossil diesel replacement and that bioethanol will be used as a gasoline replacement. Biomass-based energy sources for heat, electricity, and transportation fuels are potentially carbon dioxide neutral and recycle the same carbon atoms. Owing to widespread availability, biorenewable fuel technology will potentially result in the employment of more people than fossil-fuel-based technology.

Renewable liquid biofuels for transportation have recently attracted huge attention in different countries because of their renewability, their sustainability, their common availability, regional development, the creation of rural manufacturing jobs, the reduction of greenhouse gas emissions, and their biodegradability.

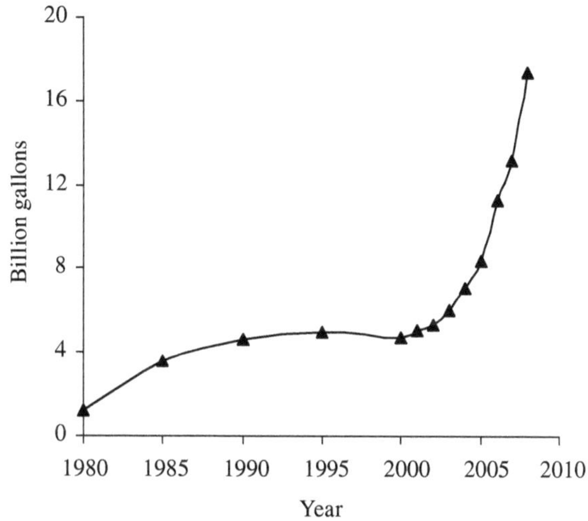

Fig. 1.7 World production of ethanol

Bioethanol can be used directly in cars designed to run on pure ethanol or blended with gasoline to make "gasohol." Anhydrous ethanol is required for blending with gasoline. No engine modification is typically needed to use the blend. Ethanol can be used as an octane-boosting, pollution-reducing additive in unleaded gasoline.

Figure 1.7 shows the world production of ethanol between 1980 and 2008 (RFA 2009). Between 1991 and 2001, world ethanol production rose from around 16 billion liters a year to 18.5 billion liters a year. From 2001 to 2007, production tripled, from 19 billion liters a year to almost 60 billion liters a year. Brazil was the world's leading ethanol producer until 2005, when US production roughly equaled Brazil's. The USA became the world's leading ethanol producer in 2006. China holds a distant but important third place in world rankings, followed by India, France, Germany, and Spain. Ethanol production by country/region in 2007 and 2008 is given in Table 1.13.

The continued increases in the price of crude oil in 2005 and 2006 resulted in a reversal of the traditional relationship between the price of biomass energy and that of crude oil, something not seen since the 1930s. As a consequence of the high prices of traded crude oil, many countries advanced their biofuel goals and, in the case of Brazil and the USA, large production gains occurred.

The other significant biofuel is biodiesel, which is currently produced from vegetable oils, animal fats, and grease by transesterification. The vegetable oils with carbon chain lengths of between 16 and 22 carbon atoms are generally in the form of triacyl glycerides, which on transesterification with methanol produce fatty acid methyl ester as the precursor to biodiesel and glycerol as a by-product. Vegetable oil (m)ethyl esters, commonly referred to as "biodiesel," are prominent candidates as alternative diesel fuels. The name biodiesel has been given to transesterified vegetable oil to describe its use as a diesel fuel. After purification of fatty acid methyl ester and testing for compliance with either EN 14214 or ASTM D6751

Table 1.13 Ethanol production by country/region in 2007 and 2008 (million gallons)

Country	2007	2008
USA	6,500	9,001
Brazil	5,020	6,473
European Union	571	734
China	487	502
Canada	212	238
Others	118	130
Thailand	80	90
Colombia	75	80
India	53	66
Australia	26	27
Turkey	16	18
Total	13,158	17,359

standards, the product can be sold as biodiesel and used as blends – typically B5 (5% biodiesel) to B20, depending on the engine warranties.

Biodiesel is a synthetic diesel-like fuel produced from vegetable oils, animal fats, or waste cooking oil. It can be used directly as a fuel, which requires some engine modifications, or can be blended with petroleum diesel and used in diesel engines with few or no modifications. At present, biodiesel accounts for less than 0.2% of the diesel consumed for transport (UN 2006). Biodiesel has become more attractive recently because of its environmental benefits. The cost of biodiesel, however, is the main obstacle to commercialization of the product. With cooking oils used as raw material, the viability of a continuous transesterification process and recovery of high-quality glycerol as a biodiesel by-product are primary options to be considered to lower the cost of biodiesel (Zhang *et al.* 2003).

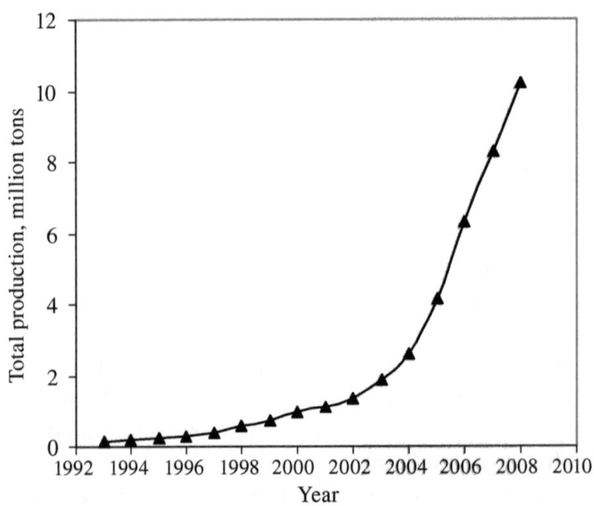

Fig. 1.8 World production of biodiesel (Demirbas 2008)

Table 1.14 Biodiesel production (2007) and production capacity (2008) of European Union countries (million tons) (EBB 2009)

Country	Production 2007	Production capacity 2008
Germany	2,890	5,302
France	872	1,980
Italy	363	1.566
Spain	168	1,267
UK	150	726
Belgium	166	665
The Netherlands	85	571
Greece	100	565
Austria	267	485
Poland	80	450
Portugal	175	406
Bulgaria	9	215
Sweden	63	212
Slovakia	46	206
Czech Republic	61	203
Hungary	7	186
Finland	39	170
Lithuania	26	147
Denmark	85	140
Estonia	0	135
Latvia	9	130
Romania	36	111
Slovenia	11	67
Ireland	3	80
Malta	1	8
Cyprus	1	6
Luxemburg	0	0
Total	5,713	16,000

Table 1.14 shows the biodiesel production (2007) and production capacity (2008) of European Union countries. Figure 1.8 shows the world production of biodiesel between 1993 and 2008.

Between 1991 and 2001, world biodiesel production grew steadily to approximately one billion liters. Most of this production was in OECD Europe and was based on virgin vegetable oils. Small plants using waste cooking oils started to be built in other OECD countries by the end of the 1990s, but the industry outside Europe remained insignificant until around 2004. Since then, governments around the world have instituted various policies to encourage development of the industry, and new capacity in North America, Southeast Asia, and Brazil has begun to come on stream at a brisk rate. As a result, between 2001 and 2007, biodiesel production grew almost 12-fold, to 11 billion liters (Demirbas 2008).

The cost of feedstock is a major economic factor in the viability of biodiesel production. Using an estimated process cost, exclusive of feedstock cost, of \$ 0.158 per liter (\$ 0.60 per gallon) for biodiesel production, and estimating a feedstock cost of \$ 0.539 per liter (\$ 2.04 per gallon) for refined soy oil, Haas *et al.*

(2006) estimated an overall cost of \$0.70 per liter (\$2.64 per gallon) for the production of soy-based biodiesel. Biodiesel from animal fat is currently the cheapest option (\$0.4 to \$0.5 per liter), whereas traditional transesterification of vegetable oil currently costs around \$0.6 to \$0.8 per liter (IEA 2007). Rough projections of the cost of biodiesel from vegetable oil and waste grease are, respectively, \$0.54 to \$0.62 per liter and \$0.34 to \$0.42 per liter. With pretax diesel priced at \$0.18 per liter in the USA and \$0.20 to \$0.24 per liter in some European countries, biodiesel is thus currently not economically feasible, and more research and technological development will be needed (Bender 1999).

The second-generation biofuels include renewable and green diesels. The former result from a technology that incorporates vegetable oils into the crude-oil-derived diesel production process to produce a renewable carbon-based diesel with no oxygen content and a very high cetane number, whereas the latter result from the production of middle distillate by means of Fischer–Tropsch catalysts, using syngas produced by the gasification of biomass. Fischer–Tropsch-like catalysts (synthol process) can also produce ethanol and mixed alcohols.

1.3.1.3 Bio-oil

The term "bio-oil" is used mainly to refer to liquid fuels. There are several reasons for bio-oils to be considered as relevant fuels by both developing and industrialized countries. They include energy security reasons, environmental concerns, foreign exchange savings, and socioeconomic issues related to the rural sector.

Bio-oils are liquid or gaseous fuels made from biomass materials, such as agricultural crops, municipal waste, and agricultural and forestry by-products, via biochemical or thermochemical processes. They can substitute conventional fuels in vehicle engines – either totally or partially in a blend.

Biomass is dried and then converted to an oily product known as bio-oil by very quick exposure to heated particles in a fluidized bed. The char and gases produced are combusted to supply heat to the reactor, whereas the product oils are cooled and condensed. The bio-oil is then shipped by truck to the hydrogen production facility. It is more economic to produce bio-oil at remote locations and then ship the oil, since the energy density of bio-oil is higher than that of biomass. For this analysis, it was assumed that the bio-oil would be produced at several smaller plants which are closer to the sources of biomass, such that lower-cost feedstocks can be obtained.

The purpose of this analysis was to assess the economic feasibility of producing hydrogen from biomass via two thermochemical processes: (1) gasification followed by reforming of the syngas (H_2 + CO) and (2) fast pyrolysis followed by reforming of the carbohydrate fraction of the bio-oil. In each process, water-gas shift is used to convert the reformed gas into hydrogen, and pressure swing adsorption is used to purify the product.

Pyrolysis/cracking is defined as the cleavage to smaller molecules by thermal energy. Hydrogen can be produced economically from woody biomass. Biomass can be thermally processed through gasification or pyrolysis to produce hydrogen. The main gaseous products from biomass are the following:

$$\text{Pyrolysis of biomass} \rightarrow H_2 + CO_2 + CO + \text{gaseous and liquid hydrocarbons} \quad (1.9)$$

$$\text{Catalytic steam reforming of biomass} \rightarrow H_2 + CO_2 + CO \quad (1.10)$$

$$\text{Fischer–Tropsch synthesis of } H_2 + CO \rightarrow \text{gaseous and liquid hydrocarbons} \quad (1.11)$$

The conventional pyrolysis of biomass is associated with the product of interest charcoal, but the fast pyrolysis is associated with the products of interest tar, at low temperature (675–775 K), and/or gas, at high temperature.

The bio-oil obtained from the fast pyrolysis of biomass has a high oxygen content. Because of the reactivity of oxygenated groups, the main problems with the oil are instability. Therefore, study of the deoxygenation of bio-oil is needed. In previous work the mechanism of hydrodeoxygenation of bio-oil in the presence of a cobalt molybdate catalyst was studied (Zhang *et al.* 2003).

The main hydrodeoxygenation reaction is represented in Eq. 1.12:

$$-(CH_2O)- + H_2 \rightarrow -(CH_2)- + H_2O. \quad (1.12)$$

This is the most important route of chemical upgrading. The reaction has strong analogies with typical refinery hydrogenations such as hydrodesulfurization and hydrodenitrification. In general, most of the hydrodeoxygenation studies have been performed using existing hydrodesulfurization catalysts (NiMo and CoMo on suitable carriers). Such catalysts need activation using a suitable sulfur source and this is a major drawback when nearly sulfur free resources such as bio-oil are used.

The primary objective of catalytic partial hydrodeoxygenation is to increase the energetic value of the oil by removing bound oxygen in the form of water. Hydrodeoxygenation of bio-oils involves treating the oils at moderate temperatures with high-pressure hydrogen in the presence of heterogeneous catalysts. The process is carried out in two distinct stages: the first stage at relatively low temperatures (525–575 K), aimed at stabilizing the bio-oil, and the second stage at higher temperatures (575–675 K), to deoxygenate the intermediate product. Different types of catalysts have been screened, ranging from conventional sulfided catalysts used in the hydrodesulfurization process (i.e., $NiMo/Al_2O_3$, $CoMo/Al_2O_3$) to novel a nonsulfided catalyst based on noble metal catalysts (i.e., Ru/Al_2O_3). The operating conditions were optimized to obtain the highest yield of a hydrocarbon-like liquid product (Huber *et al.* 2005; Metzger 2006).

It is possible to produce light alkanes by aqueous-phase reforming of biomass-derived oxygenates such as sorbitol, which can be obtained from glucose by hydrogenation. The production of alkanes from aqueous carbohydrate solutions would be advantageous because of the easy separation of the alkanes from water. Much hydrogen is needed to reduce biomass-derived oxygenates to alkanes, as shown in Eq. 1.13.

$$C_6H_{14}O_6 + 6H_2 \rightarrow C_6H_{14} + 6H_2O \quad (1.13)$$

It was shown in earlier work that hydrogen can be produced in a catalytic process at 500 K in water from biomass-derived polyols having a carbon-to-oxygen stoichiometry of 1:1, such as sorbitol, as shown in Eq. 1.14 (Metzger 2006).

$$C_6H_{14}O_6 + 6H_2O \rightarrow 13H_2 + 6CO_2 \tag{1.14}$$

The overall reaction, Eq. 1.15, is an exothermic process and converts approximately 1.5 mol of sorbitol into 1 mol of hexane.

$$19C_6O_6H_{14} \rightarrow 13C_6H_{14} + 36CO_2 + 42H_2O \tag{1.15}$$

About 95% of the heating value but only 30% of the mass of the biomass-derived reactant is retained in the product. The remaining 70% of the biomass is found as carbon dioxide and water. A maximum yield of 61% hydrogen with respect to sorbitol was obtained over Pt/Al_2O_3 under optimized reaction conditions (Metzger 2006).

1.3.1.4 Biogas

The organic fraction of almost any form of biomass, including sewage sludge, animal waste, and industrial effluents, can be broken down through anaerobic digestion into a methane and carbon dioxide mixture called as "biogas." Biogas is an environmentally friendly, clean, cheap, and versatile fuel which is produced in digesters filled with the feedstock. The digestion is allowed to continue for a period of from 10 days to a few weeks.

Biogas can be obtained from several sources. It is obtained from decomposing organic material. Biogas is composed of methane, carbon dioxide, air, ammonia, carbon monoxide, hydrogen, sulfur gases, nitrogen, and oxygen. Among its components, methane is the most important one, particularly for the combustion process in vehicle engines. Methane and carbon dioxide make up around 90% of the gas volume produced. The main constituents of landfill gas are methane and carbon dioxide, both of which are major contributors to global warming. Because of the widely varying nature of the contents of landfill sites, the constituents of landfill gases vary widely.

The total solids content in the slurry is determined by drying it in an oven at 378 K until a constant weight is obtained. The dried solid samples from the total solids determination are ignited at 1,225 K in a furnace for 7 min. The loss in weight is taken as the weight of volatile solids of the substrate slurry.

After the first 6 days of digestion, methane production from manure increases exponentially, after 16 days it reaches a plateau, and at the end of the 20th day the digestion reaches the stationary phase. For wheat straw and mixtures of manure and straw, the rates of digestion are lower than that of manure.

The maximum daily biogas production is between 4 and 6 days. During a 30-day digestion period, approximately 80–85% of the biogas is produced in the first 15–18 days. This implies that the digester retention time can be designed for 15–18 days instead of 30 days.

For the first 3 days, the methane yield is almost 0% and carbon dioxide generation is almost 100%. In this period, digestion occurs as fermentation to carbon dioxide. The yields of methane and carbon dioxide gases are 50:50 at the 11th day. At the end of the 20th day, the digestion reaches the stationary phase. The methane content of the biogas is in the range 73–79% for the runs, the remainder being principally carbon dioxide. During digestion, the volatile fatty acid concentration is lower and the pH higher. The pH of the slurry with manure increases from 6.4, initially, to 6.9–7.0 at the maximum methane production rate. The pH of the slurry with wheat straw is around 7.0–7.1 at the maximum methane production rate.

1.3.1.5 Fischer–Tropsch Liquids

Syngas (a mixture of carbon monoxide and hydrogen) produced by gasification of fossil fuels or biomass can be converted into a large number of organic compounds that are useful as chemical feedstocks, fuels, and solvents. Many of the conversion technologies were developed for coal gasification, but process economics has resulted in a shift to natural-gas-derived syngas. These conversion technologies successively apply similarly to biomass-derived biosyngas. Fischer and Tropsch first studied the conversion of syngas into larger, useful organic compounds in 1923 (Spath and Mann 2000).

Biofuels are considered to important fuels by both developing and industrialized countries owing to energy security reasons, environmental concerns, foreign exchange savings, and socioeconomic issues related to the rural sector. Catalytic conversion will be a primary tool for industry to produce valuable fuels, chemicals, and materials from biomass platform chemicals. Catalytic conversion of biomass is best developed for syngas. Economic considerations dictate that the current production of liquid fuels from syngas translates into the use of natural gas as the hydrocarbon source. Biomass is the only renewable that can meet our demand for carbon-based liquid fuels and chemicals. Biofuels as well as green motor fuels produced from biomass by Fischer–Tropsch synthesis are the most modern biomass-based transportation fuels. Green motor fuels are the renewable replacement for petroleum-based diesel. Biomass energy conversion facilities are important for obtaining bio-oil by pyrolysis.

Fischer–Tropsch synthesis is a process for producing mainly straight-chain hydrocarbons from a syngas rich in CO and H_2. The syngas must have very low tar and particulate matter content. Catalysts are usually employed. Typical operating conditions for Fischer–Tropsch synthesis are temperatures of 475–625 K and very high pressure, depending on the desired products. The product range includes light hydrocarbons such as methane (CH_4) and ethane (C_2H_6), propane (C_3H_8), butane (C_4H_{10}), gasoline (C_5–C_{12}), diesel (C_{13}–C_{22}), and light hydrocarbons and waxes (C_{23}–C_{33}). The distribution of the products depends on the catalyst and the process conditions (temperature, pressure, and residence time). The process for producing liquid fuels from biomass, which integrates biomass gasification with Fischer–Tropsch synthesis, converts a renewable feedstock into a clean fuel.

The Fischer–Tropsch synthesis is described by the equation (Schulz 1999)

$$nCO + (n + m/2)H_2 \rightarrow C_nH_m + nH_2O, \quad (1.16)$$

where n is the average length of the hydrocarbon chain and m is the number of hydrogen atoms per carbon. All reactions are exothermic and the product is a mixture of different hydrocarbons, of which paraffin and olefins are the main constituents.

The first Fischer–Tropsch synthesis plants began operation in Germany in 1938 but closed down after World War II. Then in 1955, Sasol, a world leader in the commercial production of liquid fuels and chemicals from coal and crude oil, started Sasol I in Sasolburg, South Africa. Following the success of Sasol I, Sasol II and Sasol III, located in Secunda, South Africa, came on line in 1980 and 1982, respectively. The Fischer–Tropsch synthesis is an essential step to convert carbon-containing feedstocks to liquid fuels such as diesel. Major advantages of the Fischer–Tropsch synthesis are (1) the flexibility in feedstocks (natural gas, coal, biomass), (2) the large and even sustainable resources, (3) it is ultraclean (low sulfur content) products, and (4) its suitability for converting difficult-to-process resources. A major drawback of the Fischer–Tropsch process is the polymerization-like nature of the process, yielding a wide product spectrum, ranging from compounds with low molecular mass, methane, to products with very high molecular mass, such as heavy waxes.

1.3.1.6 Biohydrogen

Hydrogen is not a primary fuel. It must be manufactured from water with either fossil or nonfossil energy sources. Widespread use of hydrogen as an energy source could have a positive impact on global climate change and improve energy efficiency and air quality. The thermochemical conversion processes, such as pyrolysis, gasification, and steam gasification, are available for converting the biomass to a more useful energy. The yield from steam gasification increases with increasing water-to-sample ratio. The yields of hydrogen from the pyrolysis and the steam gasification increase with increasing temperature. A list of some biomass material used for hydrogen production is given in Table 1.15. Hydrogen-powered fuel cells are an important enabling technology for the hydrogen future and are more-efficient alternatives to the combustion of gasoline and other fossil fuels. Hydrogen has the potential to solve two major energy problems: reducing dependence on petroleum and reducing pollution and greenhouse gas emissions.

A fuel cell is a device or an electrochemical engine that converts the energy of a fuel directly to electricity and heat without combustion. Fuel cells consist of two electrodes sandwiched around an electrolyte. When oxygen passes over one electrode and hydrogen passes over the other, electricity is generated. Fuel cells running on hydrogen derived from a renewable source would emit nothing but water vapor. Fuel cells are clean, quiet, and efficient.

Table 1.15 Some biomass material used for hydrogen production

Biomass species	Main conversion process
Bionut shell	Steam gasification
Olive husk	Pyrolysis
Tea waste	Pyrolysis
Crop straw	Pyrolysis
Black liquor	Steam gasification
Municipal solid waste	Supercritical water extraction
Crop grain residue	Supercritical fluid extraction
Pulp and paper waste	Microbial fermentation
Petroleum-based plastic waste	Supercritical fluid extraction
Manure slurry	Microbial fermentation

Hydrogen is currently more expensive than conventional energy sources. Different technologies are presently being used to produce hydrogen economically from biomass. Hydrogen can be produced by pyrolysis from biomass. It can be burned to produce heat or passed through a fuel cell to produce electricity. Biomass represents a large potential feedstock resource for environmentally clean hydrogen production. It lends itself to both biological and thermal conversion processes. In the thermal path, hydrogen can be produced in two ways: direct gasification and pyrolysis to produce liquid bio-oil, followed by steam reforming. Hydrogen can alternatively be produced by supercritical water gasification of biomass (Bryd *et al.* 2007).

Biohydrogen technology will play a major role in the future because it can utilize a renewable source of energy. Figure 1.9 shows the share of hydrogen in total automotive fuel consumption in the world as a futuristic view.

Biological generation of hydrogen (biohydrogen) technologies provide a wide range of approaches to generate hydrogen, including direct biophotolysis, indirect

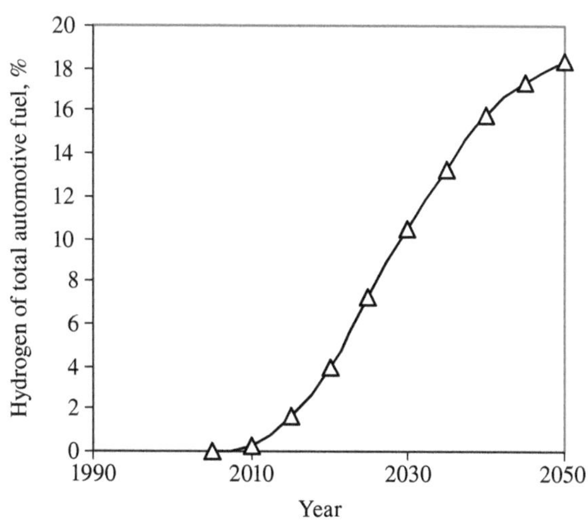

Fig. 1.9 Share of hydrogen in total automotive fuel in the future (Demirbas 2008)

biophotolysis, photofermentations, and dark fermentation. Biological hydrogen production processes are found to be more environmentally friendly and less energy intensive than thermochemical and electrochemical processes. Researchers have been investigating hydrogen production with anaerobic bacteria since the 1980s.

Three types of microorganisms are used for hydrogen generation: cyanobacteria, anaerobic bacteria, and fermentative bacteria. The cyanobacteria directly decompose water to hydrogen and oxygen in the presence of light energy by photosynthesis. Photosynthetic bacteria use organic substrates such as organic acids. Anaerobic bacteria use organic substances as the sole source of electrons and energy, converting them into hydrogen. Biohydrogen can be generated using bacteria such as clostridia by adjusting the temperature, pH, reactor hydraulic retention time, and other factors of the treatment system.

Biological hydrogen can be generated from plants by biophotolysis of water using microalgae (green algae and cyanobacteria), fermentation of organic compounds, and photodecomposition of organic compounds by photosynthetic bacteria. To produce hydrogen by fermentation of biomass, a continuous process using a nonsterile substrate with a readily available mixed microflora is desirable (Hussy *et al.* 2005). A successful biological conversion of biomass to hydrogen depends strongly on the processing of raw materials to produce feedstock which can be fermented by the microorganisms.

Hydrogen production from the bacterial fermentation of sugars has been examined in a variety of reactor systems. Hexose concentration has a greater effect on hydrogen yields than the hydraulic retention time. Flocculation was also an important factor in the performance of the reactor.

Hydrogen gas is a product of the mixed acid fermentation of *Escherichia coli*, the butylene glycol fermentation of *Aerobacter*, and the butyric acid fermentations of *Clostridium* spp. It was conducted to improve hydrogen fermentation of food waste in a leaching-bed reactor by heat-shocked anaerobic sludge, and also to investigate the effect of dilution rate on the production of hydrogen and metabolites in hydrogen fermentation.

1.3.1.7 Sugar from Biomass

Cellulose is a remarkable pure organic polymer, consisting solely of units of anhydroglucose held together in a giant straight-chain molecule. Cellulose must be hydrolyzed to glucose before fermentation to ethanol. The efficiencies of conversion of cellulose to glucose are dependent on the extent of chemical and mechanical pretreatments to structurally and chemically alter the pulp and paper mill waste. The method of pulping, the type of wood, and the use of recycled pulp and paper products can also influence the accessibility of cellulose to cellulase enzymes. Hemicelluloses (arabinoglycuronoxylan and galactoglucomammans) are related to plant gums in composition, and occur in much shorter molecular chains than cellulose. The hemicelluloses, which are present in deciduous woods chiefly as pentosans and in coniferous woods almost entirely as hexosanes, undergo thermal decomposition very readily. Hemicelluloses are derived mainly from chains of pentose sugars, and act as the cement material holding together the cellulose micelles and fiber. Cellulose is insoluble in most solvents and has a low accessibility to acid and enzymatic hydrolysis. Hemicelluloses are largely soluble in alkalis and, as such, are more easily hydrolyzed.

Hydrolysis (saccharification) breaks the hydrogen bonds in the hemicellulose and cellulose fractions and results in conversion to the sugar components: pentoses and hexoses. These sugars can then be fermented into bioethanol.

Cellulose [hexosan, $(C_6H_{10}O_5)_n$] hydrolysis produces glucose (a hexose, $C_6H_{12}O_6$). The hydrolysis of cellulose is catalyzed by mineral acids and enzymes.

Hexosan to hexose:

$$(C_6H_{10}O_5)_n + nH_2O \rightarrow nC_6H_{12}O_6 \quad (1.18)$$

Hemicellulose hydrolysis produces both hexose and pentose sugars: mannose, galactose, xylose, and arabinose, which are not all fermented with existing strains. The hemicellulose fraction typically produces a mixture of sugars including xylose, arabinose, galactose, and mannose. These are pentosans (xylose and arabinose) and hexosans (galactose and mannose). The hydrolysis of hemicelluloses is catalyzed by mineral acids and enzymes.

Pentosan to pentose:

$$(C_5H_8O_4)_m + mH_2O \rightarrow mC_5H_{10}O_5 \quad (1.19)$$

The most commonly applied methods can be classified in two groups: chemical hydrolysis (dilute and concentrated acid hydrolysis) and enzymatic hydrolysis. In addition, there are some other hydrolysis methods in which no chemicals or enzymes are applied. For instance, lignocellulose may be hydrolyzed by γ-ray or electron-beam irradiation, or microwave irradiation. However, these processes are commercially unimportant.

Both enzymatic and chemical hydrolyses require a pretreatment to increase the susceptibility of cellulosic materials. In the chemical hydrolysis, the pretreatment and the hydrolysis may be carried out in a single step. Two basic types of the acid hydrolysis process are commonly used – dilute acid and concentrated acid – each with variations.

The dilute acid process is conducted under high temperature and pressure, and has a reaction time in the range of seconds or minutes, which facilitates continuous processing. As an example, using a dilute acid process with 1% sulfuric acid in a continuous-flow reactor at a residence time of 0.22 min and a temperature of 510 K with pure cellulose provided a yield of over 50% sugars. In this case, 1,000 kg of dry wood would yield about 164 kg of pure ethanol. The combination of acid and high temperature and pressure requires special reactor materials, which can make the reactor expensive. The first reaction converts the cellulosic materials to sugars and the second reaction converts the sugars to other chemicals. Unfortunately, the conditions that cause the first reaction to occur are also the right conditions for the second reaction to occur.

The biggest advantage of dilute acid processes is their fast rate of reaction, which facilitates continuous processing. Since C_5 sugars degrade more rapidly than C_6 sugars, one way to decrease sugar degradation is to have a two-stage process. The first stage is conducted under mild process conditions to recover the C_5 sugars, whereas the second stage is conducted under harsher conditions to recover the C_6 sugars.

Hydrolysis of cellulosic materials by concentrated sulfuric or hydrochloric acids is a relatively old process. The concentrated acid process uses relatively mild temperatures, and the only pressures involved are those created by pumping materials from vessel to vessel. The reaction times are typically much longer than for the dilute acid process. This method generally uses concentrated sulfuric acid followed by a dilution with water to dissolve and hydrolyze or convert the substrate into sugar. This process provides a complete and rapid conversion of cellulose to glucose and hemicelluloses to C_5 sugars with little degradation. The critical factors needed to make this process economically viable are optimization of sugar recovery and cost-effective recovery of the acid for recycling. The solid residue from the first stage is dewatered and soaked in a 30–40% concentration of sulfuric acid for 1–4 h as a pre-cellulose-hydrolysis step. The solution is again dewatered and dried, increasing the acid concentration to about 70%. After reaction in another vessel for 1–4 h at low temperatures, the contents are separated to recover the sugar and acid. The sugar/acid solution from the second stage is recycled to the first stage to provide the acid for the first-stage hydrolysis.

The primary advantage of the concentrated acid process is the potential for high sugar recovery efficiency. The acid and sugar are separated via ion exchange and then the acid is reconcentrated via multiple-effect evaporators. The low temperatures and pressures employed allow the use of relatively low cost materials such as fiberglass tanks and piping. The low temperatures and pressures also minimize the degradation of sugars. Unfortunately, it is a relatively slow process and cost-effective acid recovery systems have been difficult to develop. Without acid recovery, large quantities of lime must be used to neutralize the acid in the sugar solution. This neutralization forms large quantities of calcium sulfate, which requires disposal and creates additional expense.

Another basic method of hydrolysis is enzymatic hydrolysis. Enzymes are naturally occurring plant proteins that cause certain chemical reactions to occur. There are two technological developments: enzymatic and direct microbial conversion methods.

The chemical pretreatment of the cellulosic biomass is necessary before enzymatic hydrolysis. The first application of enzymatic hydrolysis was used in separate hydrolysis and fermentation steps. Enzymatic hydrolysis is accomplished by cellulolytic enzymes. Different kinds of "cellulases" may be used to cleave the cellulose and hemicelluloses. A mixture of endoglucanases, exoglucanases, β-glucosidases, and cellobiohydrolases is commonly used. The endoglucanases randomly attack cellulose chains to produce polysaccharides of shorter length, whereas exoglucanases attach to the nonreducing ends of these shorter chains and remove cellobiose moieties. β-Glucosidases hydrolyze cellobiose and other oligosaccharides to glucose (Demirbas 2001).

For enzymes to work efficiently, they must obtain access to the molecules to be hydrolyzed. This requires some kind of pretreatment process to remove hemicelluloses and break down the crystalline structure of the cellulose or removal of the lignin to expose hemicelluloses and cellulose molecules.

1.3.2 Hydropower

The water in rivers and streams can be captured and turned into hydropower, also called hydroelectric power. Large-scale hydropower provides about one quarter of the world's total electricity supply, virtually all of Norway's electricity, and more than 40% of the electricity used in developing countries. In developing countries, considerable potential still exists, but large hydropower projects may face financial, environmental, and social constraints (UNDP 2000). The technically usable world potential of large-scale hydropower is estimated to be over 2,200 GW, of which only about 25% is currently exploited.

Micro hydropower systems are sometimes described as those having capacities below 100 kW, mini hydropower plants are those ranging from 100 to 1,000 kW, and small hydropower plants are those that produce from 1 to 30 MW.

Micro hydropower systems which use cross flow turbines and Pelton wheels can provide both direct mechanical energy (for crop processing) and electric energy. However, owing to design constraints, turbines up to a capacity of 30 kW are suitable for extracting mechanical energy. Of the total installed capacity of about 12 MW of micro hydropower systems, half is used solely for crop processing. The most popular micro hydropower system is the peltric set, which is an integrated Pelton turbine and electricity generation unit with an average capacity of 1 kW.

There is no universal consensus on the definition of small-scale hydropower. Some countries in the European Union such as Portugal, Spain, Ireland, Greece, and Belgium define 10 MW as the upper limit for installed capacity. In Italy the limit is 3 MW, in France 8 MW, in the UK 5 MW, in Canada 20–25 MW, and in the USA 30 MW; however, a value of up to 10-MW total capacity is becoming generally accepted as small-scale hydropower in the rest of the world. If the total installed capacity of any hydropower system is more than 10 MW, it is generally accepted that it is a large hydropower system (Demirbas 2006). Small-scale hydropower can be further subdivided into mini hydropower, usually defined as less than 1,000 kW, and micro hydropower, which is less than 100 kW. The definitions of micro hydropower and small-scale hydropower vary from country to country. Small-scale hydropower is one of the most valuable energies to be offered for electrification of rural communities. The growth of small-scale hydroelectricity serves to decrease the difference in decentralized production between private sector and municipal activity production. Small-scale hydropower systems use the energy from flowing or running water and convert it to electric energy. The potential for small-scale hydropower systems depends on the availability of

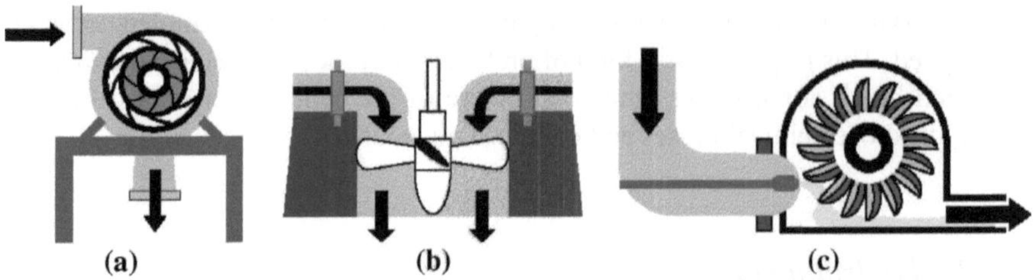

Fig. 1.10 Main types of water turbines: (a) Francis, (b) Kaplan, and (c) Pelton

water flow where the resource exists. A well-designed small-scale hydropower system can be established anywhere, it can fit in with its surroundings, and it will have minimal negative impacts on the environment. Small-scale hydropower systems allow self-sufficiency to be achieved by using scarce natural water resources. These systems provide low-cost energy production and can be used in many developing countries.

A water power plant is, in general, a highly effective energy-conversion system. There is no pollution of the environment, but objections are raised relating to the flooding of valuable real estate and scenic areas. Whether a particular hydroelectric installation is economically competitive with a fossil fuel power plant will depend upon a number of factors, in particular, fuel and construction costs. As far as nonfossil energy is concerned, hydropower and nuclear power resources are the principal assets, owing to their high production potential and their economic efficiency.

There are two types of turbines: reaction turbines and impulse turbines. Figure 1.10 shows the main types of water turbines used in hydropower applications. In reaction turbines, such as Francis and Kaplan turbines, the pressure of water applies a force onto the face of the runner blades, which decreases as it proceeds through the turbine. Francis turbines are generally used in a head range of 5–250 m and can be designed with either a vertical or a horizontal shaft. Kaplan turbines are axial-flow reaction turbines, generally used for low heads.

In impulse turbines, water pressure is converted into kinetic energy in the form of a high-speed jet that strikes buckets mounted on the periphery of the runner. The most common impulse type is the Pelton turbine. It is generally used in installations with a head of 50 to several hundred meters. By adjustment of the flow through the nozzle, a Pelton turbine can operate at high efficiency over a wide range of head and flow conditions.

Typically, larger turbines have higher efficiencies. For example, the efficiency is usually above 90% for turbines producing several hundred kilowatts or more, whereas the efficiency of a micro hydropower turbine of 10 kW is likely to be on the order of 60–80%.

Two main types of generators are used in the small-scale hydropower industry: synchronous and asynchronous generators. Both generator types are very well known throughout the industry and have been steadily improved to meet the needs and demands of the hydropower sector.

Lifting water from the source to a higher location can usually be carried out through a number of potential water-lifting options, depending on the particular site conditions. One means of lifting water is the hydraulic ram (hydram) pump. Hydram pumps are water-lifting or water-pumping devices that are powered by filling water. They have been used for over two centuries in many parts of the world. The pump works by using the energy of a large amount of water to lift a small amount of that water to a much greater height. Wherever a fall of water can be obtained, a hydram pump can be used as a comparatively cheap, simple, and reliable means of raising water to considerable heights. A hydram pump can be used for lifting water from a source lying at a lower elevation to a point of use located at a higher elevation for domestic use, drinking, cooking, and washing, and irrigation of small areas, gardens, and orchards. The main and unique advantage of a hydram pump is that with a continuous flow of water, it operates automatically and continuously with no other external energy source – be it electricity or hydrocarbon fuel. It uses a renewable energy source (stream of water) and hence ensures low running costs (Demirbas 2006).

Energy that is stored in the gravitational field is called gravitational potential energy, or potential energy due to gravity. If the object is being lifted at constant velocity, then it is not accelerating, and the net force on it is zero. When you lift something at a constant velocity, the force that you lift with equals the weight of the object. So, the work done lifting an object is equal to its mass times the acceleration due to gravity times the height of the lift. As the object falls, it travels faster and faster, and thus picks up more and more kinetic energy. This increase in kinetic energy during the fall is due to the drop in gravitational potential energy during the fall. The gravitational potential energy becomes the kinetic energy of the falling object. The water above receives energy as it falls down the short waterfall. This energy is stored as potential energy in the gravitational field of the earth and is released as the water drops. This energy is expressed as the kinetic energy of the water.

The efficiency of small-scale hydropower depends mainly on the performance of the turbine. Today, generators commonly have efficiency rates of 98–99%. As a general rule, larger and newer plants have higher efficiencies of up to 90%. The efficiency can be as low as 60% for old and smaller plants. Hydropower is the most efficient way to generate electricity. Modern hydro turbines can convert as much as 90% of the available energy into electricity. The best fossil fuel plants are only about 50% efficient.

Hydropower provides unique benefits, rarely found with other sources of energy. These benefits can be attributed to the electricity itself, or to side benefits, often associated with reservoir development. Principles of sustainable development of particular relevance to energy provision include improving the well-being of entire populations, ensuring that development is people-centered, participatory, and equitable; integrating environmental concerns into decision-making processes; and taking into account the full range of costs and benefits of development.

Investment costs for small hydropower plants vary according to site-specific and local characteristics. The most important system and cost elements are: (1)

civil engineering, (2) equipment, and (3) turbines. The electric generator represents less than 5% of the total cost of a power plant and the efficiency of generators for new plants is already close to 100%. Yet standardization of generator equipment for small-scale hydropower could further reduce installation and maintenance costs.

Despite the recent debates, few would deny that the net environmental benefits of hydropower are far superior to those of fossil-based generation. Hydroelectricity is produced for an average of 0.85 cents per kilowatt-hour. In comparison with hydropower, thermal plants take less time to design, obtain approval, build, and recover investment.

The remaining economically exploitable potential is about 5,400 TWh/year. An investment of at least US $ 1,500 billion would be necessary to realize such a program. The mean level of hydropower plant capacity is in the range 50–100 MW; some 20,000 plants would need to be built. To implement a plant construction program of this magnitude, a great deal of work (technical, financial, and political) would need to be accomplished by all those involved, particularly in Asia, South America, and Africa (Demirbas 2006).

1.3.3 Geothermal Energy

As an energy source, geothermal energy has come of age. It is clean, cheap, and renewable, and can be utilized in various forms, such as space heating and domestic hot water supply, CO_2 and dry-ice production process, heat pumps, greenhouse heating, swimming and balneology (therapeutic baths), industrial processes and

Table 1.16 World's top countries using geothermal energy in direct uses

Country	Installed capacity (MW thermal)	Production (GWh/year)
China	2,282	10,531
Japan	1,167	7,482
USA	3,766	5,640
Iceland	1,469	5,603
Turkey	820	4,377
New Zealand	308	1,967
Georgia	250	1,752
Russia	308	1,707
France	326	1,360
Sweden	377	1,147
Hungary	473	1,135
Mexico	164	1,089
Italy	326	1,048
Romania	152	797
Switzerland	547	663

electricity generation. Electricity generation is improving faster in geothermal-energy-rich countries. Geothermal energy for electricity generation has been produced commercially since 1913, and for four decades on the scale of hundreds of megawatts both for electricity generation and for direct use. The main types of direct use are bathing, swimming and balneology (42%), space heating (35%), greenhouses (9%), fish farming (6%), and industry (6%) (Fridleifsson 2001). The utilization has increased rapidly during the last three decades. The technology, reliability, economics, and environmental acceptability of direct use of geothermal energy have been demonstrated throughout the world. In 2000, geothermal resources were identified in over 80 countries and there are quantified records of geothermal utilization in 58 countries (Fridleifsson 2001). The world's top countries using geothermal energy for direct use are given in Table 1.16. Electricity is produced with geothermal steams in 21 countries spread over all continents. Low-temperature geothermal energy is exploited in many countries to generate heat, with an estimated capacity of about 10,000 MW thermal.

1.3.4 Wind Energy

Wind power has long been used for grain-milling and water-pumping applications. Significant technical progress since the 1980s, however, driven by advances in aerodynamics, materials, design, controls, and computing power, has led to economically competitive electric energy production from wind turbines. Technology development, favorable economic incentives (due to its early development status and environmental benefits), and increasing costs of power from traditional fossil sources have led to significant worldwide sales growth since the early 1980s (Kutz 2007).

Wind energy is renewable, mostly distributed generation characterized by large variations in production. The wind energy sector is one of the fastest-growing energy sectors in the world. From 1991 until the end of 2002, global installed capacity increased from about 2 GW to over 31 GW, with an average annual growth rate of about 26%. During this period, both the prices of wind turbines and the cost of wind-generated electricity were reduced. The production of energy from wind power would need to be increased by 1% to compensate for the losses of hydropower production. The energy produced from wind power, on an hourly level for 1–2 days ahead, is more difficult to predict than that produced from other sources (Demirbas 2006).

Wind energy is an indigenous supply permanently available in virtually every nation of the world. It is abundant, renewable, widely distributed, clean, and will mitigate the greenhouse effect if it is used to replace fossil-fuel-derived electricity. Wind energy has limitations based on geography and meteorology, plus there may be political or environmental problems (e.g., dead birds) with installing turbines (Garg and Datta 1998). On the other hand, wind can cause air pollution by degradation and distribution of pieces of pollutants such as waste paper and straw.

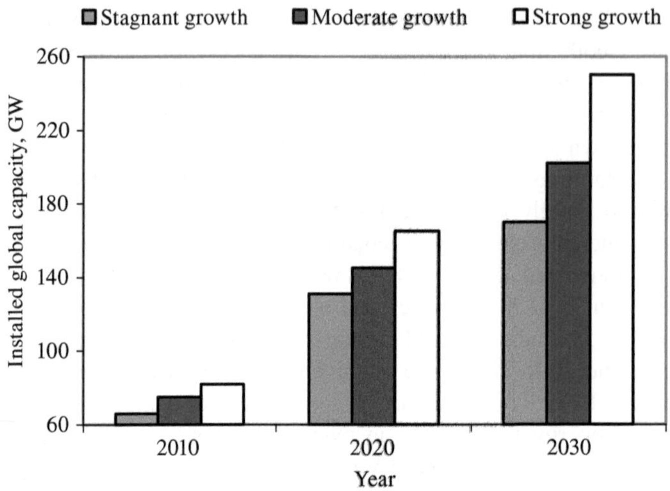

Fig. 1.11 Scenarios for global installed wind power (IEA 2006)

A detailed analysis of the technical, economic, and regulatory issues of wind power can be found in the European Wind Energy Association report "Large scale integration of wind energy in the European power supply: analysis, issues and recommendations" published in December 2005. In 2005, worldwide capacity of wind-powered generators was 58,982 MW. Although wind power currently produces less than 1% of electricity used worldwide, it accounts for 23% of electricity use in Denmark, 4.3% in Germany, and approximately 8% in Spain. Globally, wind power electricity generation more than quadrupled between 1999 and 2005 according to the European Wind Energy Association (2005).

Figure 1.11 shows the growth scenarios for global installed wind power. In 2004, the IEA reference scenario projections for stagnant growth of wind energy were 66 GW in 2010, 131 GW in 2020, and 170 GW in 2030. For moderate growth, the IEA reference scenario projections for wind energy were 75 GW in 2010, 145 GW in 2020, and 202 GW in 2030. The IEA advanced strong growth scenario projected a wind energy market of 82 GW in 2010, 165 GW in 2020 and 250 GW in 2030.

1.3.5 Solar Energy

Solar energy is defined as that radiant energy transmitted by the sun and intercepted by Earth. It is transmitted through space to Earth as electromagnetic radiation with wavelengths ranging between 0.20 and 15 μm. The availability of solar flux for terrestrial applications varies with season, time of day, location, and collecting surface orientation. In this section we shall treat these matters analytically (Kutz 2007).

Energy is obtained from solar heating, solar home heating, solar dryers, solar cookers, solar water heating, solar photovoltaic systems (converting sunlight directly into electricity), and solar thermal electric power (when the sun's energy is concentrated to heat water and produce steam, which is used to produce electricity).

Solar dryers are used for drying fruits and spices. The three most popular types of solar dryer are the box type, the cabinet type, and the tunnel type. The box type uses direct heat for dehydration. In cabinet-type dryers, air heated by the collector dehydrates the food product, whereas in the tunnel type, forced air circulation is used to distribute heat for dehydration. Cabinet- and tunnel-type dryers yield dried products of high quality, but they are very bulky and costly compared with the box-type dryers. Of about 800 dryers in use so far, 760 are of the box type (Pokharel 2003).

One of the most abundant energy resources on the surface of the earth is sunlight. Today, solar energy has a tiny contribution to the world's total primary energy supply of less than 1%. Photovoltaic systems, other than solar home heating systems, are used for communication, water pumping for drinking and irrigation, and electricity generation. The total installed capacity of such systems is estimated at about 1,000 kW. A solar home heating system is a solar photovoltaic system with a maximum capacity of 40 W. These systems are installed and managed by a household or a small community (Garg and Datta 1998).

The major component of any solar system is the solar collector. Solar energy collectors are a special kind of heat exchanger that transforms solar radiation energy to the internal energy of the transport medium. A historical introduction to the uses of solar energy was attempted followed by a description of the various types of collectors, including flat-plate, compound parabolic, evacuated tube, parabolic trough, Fresnel lens, parabolic dish, and heliostat field collectors (Kalogirou 2004). Solar thermal electricity systems utilize solar radiation to generate electricity by the photothermal conversion method.

Solar-based chemical processes utilize solar energy to drive chemical changes. These processes offset energy that would otherwise be required from an alternative source and can serve as a method of converting solar energy into a storable and transportable fuel. Solar chemical reactions are diverse, but can be generically described as either thermochemical or photochemical. A more conventional approach uses process heat from a solar concentrator to drive the steam reformation of natural gas, thereby increasing the overall hydrogen yield. Concentrators can be used in the gasification of feedstocks such as coal, woody biomass, municipal solid waste, and agricultural residues. The resulting hydrocarbons can be used to synthesize so-called sunfuels. Concentrating solar technologies can also be used in the production of industrial chemicals. Photoelectrochemical cells consist of a semiconductor, typically titanium dioxide or related titanates, immersed in an electrolyte. A photogalvanic device is a type of battery in which the cell solution forms energy-rich chemical intermediates when illuminated. These chemical intermediates then react at the electrodes to produce an electric potential.

1.3.6 Other Renewable Energy Sources

Wave energy, tidal energy, and ocean thermal energy conversion (OTEC) are the other noncombustible renewable energy sources. Tides and waves are marine energy sources and OTEC is a marine energy technology. The world wave resource is between 200 and 5,000 GW, mostly in offshore locations (Garg and Datta 1998). Wave energy converters fixed to the shoreline are likely to be the first to be fully developed and deployed, but waves are typically 2–3 times more powerful in deep offshore waters than at the shoreline. Wave energy can be harnessed in coastal areas, close to the shore. The first patent for a wave energy device was filed in Paris in 1799, and by 1973 there were 340 British patents for wave energy devices. By comparison with wind energy and photovoltaic energy technologies, wave energy and tidal stream technologies are very much in their infancy. Currently, wave energy devices with a capacity of around 1 MW are installed worldwide, mainly in demonstration projects.

OTEC is an energy technology that converts solar radiation to electric power. OTEC systems use the ocean's natural thermal gradient to drive a power-producing cycle. As long as the temperature between the warm surface water and the cold deep water differs by about 20 K, an OTEC system can produce a significant amount of power. The oceans are thus a vast renewable resource, with the potential to help us produce billions of watts of electric power.

1.3.7 Municipal Solid Waste and Landfill Gas

The generation of municipal solid waste has increased in parallel with rapid industrialization. Approximately 16% of all discarded municipal solid waste is incinerated; the remainder is disposed of in landfills. Disposal of municipal solid waste in sanitary landfills is usually associated with soil, surface water, and groundwater contamination when the landfill is not properly constructed. The flow rate and composition of leachate vary from site to site, seasonally at each site, and depend on the age of the landfill. The processing of municipal solid waste (e.g., landfill, incineration, aerobic composting) has many advantages and limitations. Greenhouse gas emissions can be reduced by the uncontrolled release of methane from improperly disposed of organic waste in a large landfill.

A typical analysis of raw landfill gas is given in Table 1.17. Methane and carbon dioxide make up around 90% of the gas volume produced, and they are both major contributors to global warming. Because of the widely varying nature of the contents of landfill sites, the constituents of landfill gases vary widely (Demirbas 2008).

Decomposition in landfills occurs in a series of stages, each of which is characterized by the increase or decrease of specific bacterial populations and the formation and utilization of certain metabolic products. The first stage of decomposition, which usually lasts less than 1 week, is characterized by the removal of oxygen

Table 1.17 Typical analysis of raw landfill gas

Component	Chemical formula	Content
Methane	CH_4	40–60 vol%
Carbon dioxide	CO_2	20–40 vol%
Nitrogen	N_2	2–20 vol%
Oxygen	O_2	<1 vol%
Heavier hydrocarbons	C_nH_{2n+2}	<1 vol%
Hydrogen sulfide	H_2S	40–100 ppm
Complex organics	–	1,000–2,000 ppm

from the waste by aerobic bacteria. In the second stage, which has been termed "the anaerobic acid stage," a diverse population of hydrolytic and fermentative bacteria hydrolyze polymers, such as cellulose, hemicellulose, proteins, and lipids, into soluble sugars, amino acids, long-chain carboxylic acids, and glycerol. The main components of landfill gas are by-products of the decomposition of organic material, usually in the form of domestic waste, by the action of naturally occurring bacteria under anaerobic conditions. Figure 1.12 shows the production of biogas components with time in a landfill.

The methods developed for treatment of landfill leachates can be classified as physical, chemical, and biological and are usually used in combination to improve the treatment efficiency. Biological treatment methods used for leachate treatment can be classified as aerobic, anaerobic, and anoxic processes and are widely used for the removal of biodegradable compounds. Biological treatment of landfill leachate usually results in low nutrient removals because of high chemical oxygen demand (COD), high ammonium nitrogen content, and the presence of toxic com-

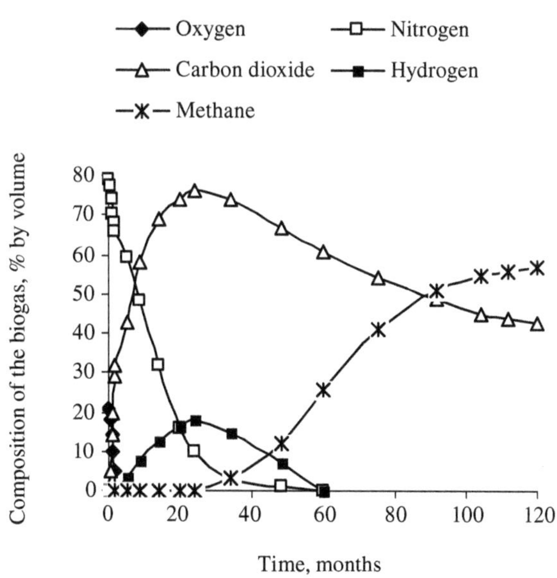

Fig. 1.12 Production of biogas components with time in a landfill

pounds such as heavy metals. Landfill leachate obtained from a solid waste landfill area contained high COD and ammonium ions; direct biological treatment resulted in low COD and ammonium removal. Several anaerobic and aerobic treatment systems have been studied for landfill leachate. Leachates contain non-biodegradable substrates which are not removed by biological treatment alone and an increase of leachate input may cause reduction of substrate removal (Cecen *et al.* 2003). Raw landfill leachate was subjected to pretreatment by coagulation–flocculation and air stripping of ammonia before biological treatment (Kargi and Pamokoglu 2004). Natural zeolite and bentonite can be utilized as a novel landfill liner material (Kayabali 1997).

1.4 Nuclear Fuel Sources

Fissile materials are those that are fissionable by neutrons with zero kinetic energy. In nuclear engineering, a fissile material is one that is capable of sustaining a nuclear fission chain reaction. Nuclear power reactors are mainly fueled by uranium, the heaviest element that occurs in nature in more than trace quantities. The principal nuclear energy sources are ^{235}U, ^{239}Pu, ^{233}U, and thorium.

The main fissile energy sources are uranium and thorium. Nuclear power plants are based on uranium mined in surface mines or obtained by *in situ* leaching. Nuclear energy has been used to produce electricity for more than half a century. Nuclear energy accounts for 6% of energy and 16% of electricity worldwide and for 23% of electricity in OECD countries (UNDP 2000). OECD countries produce almost 55% of the world's uranium. Fossil fuels – oil, coal, and gas – currently provide more than two thirds of the world's energy and electricity, but also produce the greenhouse gases largely responsible for global warming.

Nuclear energy has certain clear advantages in that it produces heat and electricity without emitting carbon dioxide into the atmosphere at the power plant level, and fuel supplies are not in danger of being exhausted. Nuclear power highly contributes to reducing global warming. For plants powered by natural gas or coal, the initial investment is lower but the fuel costs are higher. Nuclear sources of energy have high investment and low production costs per unit of power produced.

The essence of a conventional nuclear reactor is the controlled fission chain reaction of ^{235}U and ^{239}Pu. This produces heat, which is used to make steam which drives a turbine. The chain reaction depends on having a surplus of neutrons to keep it going. Work has been done in developing thorium cycle converter reactor systems. Several prototypes, including the high-temperature gas-cooled reactor and the molten salt converter reactor experiment, have operated. Whereas the high-temperature gas-cooled reactors are efficient, they are not expected to become important commercially for many years because of certain operating difficulties. Thorium is recovered commercially from the mineral monazite, which contains from 3 to 9% ThO_2 along with rare-earth minerals. Much of the internal heat the earth produces has been attributed to thorium and uranium.

In the long term, thorium may well provide a means of resource extension, in parallel with fast-reactor developments rather than as a substitute for them. Meanwhile, the most valuable contribution to industry that it can make appears likely to be in extending fuel irradiation where recycling of recovered fissile material is considered impracticable or undesirable.

For many years there has been interest in utilizing thorium (^{232}Th) as a nuclear fuel since it is 3 times as abundant in the earth's crust as uranium. Also, all of the mined thorium is potentially usable in a reactor, compared with 0.7% of natural uranium, so some 40 times the amount of energy per unit mass might be available. A thorium reactor would work by having ^{232}Th capture a neutron to become ^{233}Th, which decays to ^{233}U, which undergoes fission. The problem is that insufficient neutrons are generated to keep the reaction going.

^{232}Th goes through six α and four β decay steps before becoming the stable isotope ^{208}Pb. ^{232}Th is sufficiently radioactive to fog a photographic plate in a few hours. Thorium disintegrates with the production of "thoron" (^{220}Rn), which is an α emitter and is a radiation hazard. Good ventilation of areas where thorium is stored or handled is therefore essential.

Several methods are available for producing thorium metal: it can be obtained by reducing thorium oxide with calcium, by electrolysis of anhydrous thorium chloride in a fused mixture of sodium and potassium chlorides, by calcium reduction of thorium tetrachloride mixed with anhydrous zinc chloride, and by reduction of thorium tetrachloride with an alkali metal.

Comparison of thorium with uranium for nuclear fuel:

- Weapons-grade fissionable material (^{233}U) is harder to retrieve safely and clandestinely from the thorium reactor than plutonium is from the uranium breeder reactor.
- Thorium produces 10–10,000 times less long-lived radioactive waste than uranium or plutonium reactors.
- Thorium comes out of the ground as a 100% pure, usable isotope, which does not require enrichment, whereas natural uranium contains only 0.7% fissionable ^{235}U.
- Because thorium does not sustain a chain reaction, fission stops by default if we stop priming it, and a runaway chain reaction accident is improbable.

1.5 Summary

Global energy sources are classified into two groups: fossil and renewable. Primary energy sources can be divided into nonrenewables and renewables. Nonrenewable energy resources include coal, petrol, gas, gas hydrate, and fissile material, whereas biomass, hydro, geothermal, solar, and wind energy are renewable energy sources.

Methane hydrates are the world's largest source of fossil fuel. Large fields of methane hydrates are scattered throughout the world's oceans and are thought to contain about as much energy as all other forms of fossil fuel combined. This unconventional hydrocarbon energy source has remained untapped, however, because traditional sources are still plentiful and less expensive to develop. Nevertheless, scientists have recently been taking a closer look at hydrates, and not only as a possible energy source. A widely quoted US Geological Survey estimate predicts that there is twice as much organic carbon in gas hydrates than in all recoverable and unrecoverable conventional fossil fuel resources, including natural gas, coal, and oil.

Renewable energy refers to primary energies that are continuously generated by solar, geothermal, and tidal processes. The energy produced within the sun is responsible for a multitude of other renewable energies (such as wind and hydropower) as well as renewable energy carriers (such as solid or liquid biofuels).

References

Bender M (1999) Economic feasibility review for community-scale farmer cooperatives for biodiesel. Bioresour Technol 70:81–87

Bushnell DJ, Haluzok C, Dadkhah-Nikoo A (1989) Biomass fuel characterization testing and evaluating the combustion characteristics of selected biomass fuels. Bonneville Power Administration, Corvallis

Byrd AJ, Pant KK, Gupta RB (2007) Hydrogen production from glucose using Ru/Al_2O_3 catalyst in supercritical water. Ind Eng Chem Res 46:3574–3579

Cecen F, Erdincler A, Kilic E (2003) Effect of powdered activated carbon addition on sludge dewaterability and substrate removal in landfill leachate treatment. Adv Environ Res 7:707–713

COSO (2007) Report to the energy watch group. EWG series no 3/2007. http://www.energywatchgroup.org/fileadmin/global/pdf/EWG_Oilreport_10-2007.pdf

Demirbas A (1991) Fatty and resin acids recovered from spruce wood by supercritical acetone extraction. Holzforschung 45:337–339

Demirbas A (2000) Mechanisms of liquefaction and pyrolysis reactions of biomass. Energy Convers Manage 41:633–646

Demirbas A (2001) Biomass resource facilities and biomass conversion processing for fuel and chemicals. Energy Convers Manage 42:1357–1378

Demirbas A (2004) Combustion characteristics of different biomass fuels. Prog Energ Combus Sci 30:219–230

Demirbas A (2006) Energy priorities and new energy strategies. Energy Educ Sci Technol 16:53–109

Demirbas A (2007) Utilization of coals as sources of chemicals. Energy Sources 29:677–684

Demirbas A (2008) Biodiesel: a realistic fuel alternative for diesel engines. Springer, London

Demirbas A (2009) Biofuels: securing the planet's future energy needs. Springer, London

Dorf RC (1977) Energy resources and policy. Addison-Wesley, Los Angeles

Dry ME (1999) Fischer–Tropsch reactions and the environment. Appl Catal A Gen 189:185–190

Durgun O, Sahin Z (2007) Theoretical investigations of effects of light fuel fumigation on diesel engine performance and emissions. Energy Convers Manage 48:1952–1964

EBB (2009) European biodiesel statistics. European Biodiesel Board, Brussels

European Wind Energy Association (2005) Large scale integration of wind energy in the European power supply: analysis, issues and recommendations

Fridleifsson IB (2001) Geothermal energy for the benefit of the people. Renew Sustain Energy Rev 5:299–312

Garg HP, Datta G (1998) Global status on renewable energy. In: Solar energy heating and cooling methods in building, international workshop: Iran University of Science and Technology, 19–20 May
Glasser WG, Sarkanen S (eds) (1989) Lignin: properties and materials. American Chemical Society, Washington
Hacisalihoglu B, Demirbas AH, Hacisalihoglu S (2008) Hydrogen from gas hydrate and hydrogen sulfide in the Black Sea. Energy Educ Sci Technol 21:109–115
Haas MJ, McAloon AJ, Yee WJ, Foglia TA (2006) A process model to estimate biodiesel production costs. Bioresour Technol 97:671–678
Huber GW, Chheda JN, Barrett JA (2005) Dumesic, production of liquid alkanes by aqueous-phase processing of biomass-derived carbohydrates. Science 308:1446–1450
Hussy I, Hawkes FR, Dinsdale R, Hawkes DL (2005) Continuous fermentative hydrogen production from sucrose and sugarbeet. Int J Hydrogen Energy 30:471–483
IEA (2006) Reference scenario projections. IEA, Paris
IEA (2007) Key world energy statistics. IEA, Paris. http://www.iea.org/Textbase/nppdf/free/2007/key_stats_2007.pdf
Kalogirou SA (2004) Solar thermal collectors and applications. Prog Energy Combust Sci 30:231–295
Kargi F, Pamukoglu MY (2004) Adsorbent supplemented biological treatment of pre-treated landfill leachate by fed-batch operation. Bioresour Technol 94:285–291
Kayabali K (1997) Engineering aspects of a novel landfill liner material: bentonite-amended natural zeolite. Eng Geol 46:105–114
Kutz M (ed) (2007) Environmentally conscious alternative energy production. Wiley, Hoboken
Lee S-Y, Holder GD (2001) Methane hydrates potential as a future energy source. Fuel Proc Technol 71:181–186
Metzger JO (2006) Production of liquid hydrocarbons from biomass. Angew Chem Int Ed 45:696–698
Molina Grima E, Acién Fernández FG, García Camacho F, Chisti Y (1999) Photobioreactors: light regime, mass transfer, and scaleup. J Biotechnol 70:231–247
Pokharel S (2003) Promotional issues on alternative energy technologies in Nepal. Energy Policy 31:307–318
Ragland KW, Aerts DJ, Baker AJ (1991) Properties of wood for combustion analysis. Bioresour Technol 37:161–168
RFA (2009) Ethanol industry statistics, Renewable Fuels Association, Washington
Sarkanen KV, Ludwig CH (eds) (1971) Lignins: occurrence, formation, structure and reactions. Wiley, New York
Schobert HH, Song C (2002) Chemicals and materials from coal in the 21st century. Fuel 81:15–32
Schulz H (1999) Short history and present trends of FT synthesis. Appl Catal A Gen 186:1–16
Spath PL, Mann MK (2000) Life cycle assessment of hydrogen production via natural gas steam reforming. TP-570-27637. National Renewable Energy Laboratory, Golden
Theander O (1985) In: Overand RP, Mile TA, Mudge LK (eds) Fundamentals of thermochemical biomass conversion. Elsevier, New York
Tillman DA (2000) Biomass cofiring: the technology, the experience, the combustion consequences. Biomass Bioenergy 19:365–384
UN (2006) International conference for renewable energies, Bonn, 10 June 2006
UNDP (2000) World energy assessment 2000–energy and the challenge of sustainability. United Nations Development Programme, New York
Zhang Y, Dub MA, McLean DD, Kates M (2003) Biodiesel production from waste cooking oil: 2. Economic assessment and sensitivity analysis. Bioresour Technol 90:229–240

Chapter 2
Natural Gas

2.1 Introduction

Natural gas is the fastest-growing primary energy source in the International Energy Outlook 2003 (IEO 2003) forecast. Natural gas consumption is projected to nearly double between 2001 and 2025, with the most robust growth in demand expected among the developing nations. Because it is a cleaner fuel than oil or coal and not as controversial as nuclear power, natural gas is expected to be the fuel of choice for many countries in the future.

Natural gas combustion is clean and emits less CO_2 than all other petroleum-derivate fuels, which makes it makes favorable in terms of the greenhouse effect. Natural gas is used across all sectors, in varying amounts, including in industrial, residential, electricity generation, commercial, and transportation sectors.

Natural gas consists of 85–95% methane, which is the simplest hydrocarbon. Natural gas is the cleanest-burning alternative fuel. Exhaust emissions from natural gas vehicles (NGVs) are much lower than those from gasoline-powered vehicles. Combustion of natural gas results in the production of less carbon dioxide compared with the combustion of diesel, but the loss in efficiency when the Otto process is used means that carbon dioxide emissions increase. All in all, combustion of natural gas in a gasoline engine gives rise to about as much carbon dioxide as the combustion of diesel in a diesel engine, measured in units of energy.

Natural gas is one of the most widely used forms of energy today. Natural gas is a cleaner-burning fuel than either gasoline or diesel. However, natural gas also contains active compounds, such as sulfur, and inert compounds, such as nitrogen and carbon dioxide. Natural gas has a high octane number (110–130) and therefore has potential for use in a high-compression engine. It is a fossil fuel and is extracted from gas sources in bedrock. It is less dense than air and a carries a certain risk of explosion. Natural gas can be stored in compressed form and is then known as compressed natural gas (CNG), or at a low temperature (111 K) in liquid form, known as liquefied natural gas (LNG). LNG has been considered as an option for fueling automobiles.

A. Demirbas, *Methane Gas Hydrate*,

Table 2.1 Typical combustion properties of natural gas

Ignition point	876 K
Flammability limits	4–16 vol% (in air)
Theoretical flame temperature (stoichiometric air-to-fuel ratio)	2,233 K
Maximum flame velocity	0.3 m/s
Specific gravity (relative density)	0.583
Water vapor content	16–32 mg/m^3
Sulfur content	5.5 mg/m^3
Higher heating value (dry basis)	36.0–40.2 MJ/m^3

The composition of pipeline natural gas varies depending on the source and the processing of the gas. Typically, natural gas consists of over 90% methane and small amounts of ethane and other hydrocarbons. It may also contain nitrogen, carbon dioxide, and trace amounts of water vapor. Natural gas at ambient temperatures and pressures is a gas. As is typical of gases, it has a very low energy density compared with other fuels. The typical combustion properties of natural gas are gives in Table 2.1. On average, 0.921 m^3 of natural gas has the same energy content as 1.12 L of gasoline. This makes use of natural gas as a transportation fuel at ambient temperatures and pressures unfeasible. To use natural gas as a transportation fuel, it must be either compressed or liquefied to increase its volumetric energy density (Demirbas 2002).

The vapors of many liquid fuels such as gasoline and liquefied petroleum gas (LPG) are denser than air. As they evaporate, gasoline and LPG vapors tend to accumulate around the source and pose an explosion hazard. Natural gas requires a greater concentration in air and a higher temperature to ignite than gasoline. In the event of a leak or accident, natural gas will ignite at 811 K and between 5 and 15% concentration in air. Gasoline will ignite at only 533 K and 1.5% concentration in air.

Ventilation must be provided in CNG vehicle maintenance garages and vehicle storage buildings through ceiling exhaust systems to prevent hazardous CNG accumulations. In vehicle operation, as the pressure regulators reduce the CNG pressure, the temperature will drop, causing water vapor in the natural gas to condense. The condensed water will restrict or block fuel flow. CNG vehicle fueling stations normally dehydrate the natural gas to prevent water condensation.

Use of CNG as a transport-sector fuel requires investment in a dramatically different fuel supply system. Gas has to be taken from a gas distribution system, usually at a pressure of 0.3–1 MPa, and compressed to 20 MPa into CNG tanks of vehicles. Without intermediate storage at the CNG vehicle filling station, a slow-fill system is used that only has compressors. About 8 h is required for refueling the vehicles. For a more rapid fill, but usually no faster than 5 min, larger compressors and intermediate storage are required.

Over the next 20 years, the role of natural gas in global energy consumption will increase substantially. The speed of the transition to natural gas will be driven by environmental constraints, increased demand, and new technologies. A potential source of natural gas lies in the enormous worldwide gas hydrate reserves. It was estimated that the size of this resource ranges up to 20,000 trillion cubic meters (Kvenvolden 1993). These deposits can cause problems and safety concerns relating to drilling, production of oil and gas, and building and operation of pipelines. Naturally occurring gas hydrates are normally found at the seafloor or in shallow sediments where the pressures and temperatures are conducive to hydrate formation.

2.2 Definition and History

Natural gas is formed deep underground, usually in areas around coal and oil. It is composed primarily of methane. It does, however, contain other chemical species, such as butane and propane. If the mixture is composed only of these species, it is called dry natural gas, as there will be no liquid components at standard pressure and temperature. Natural gas might also contain nonhydrocarbon compounds, such as water vapor, carbon dioxide, and hydrogen sulfide.

Natural gas is a mixture of lightweight alkanes. Natural gas contains methane (CH_4), ethane (C_2H_6), propane (C_3H_8), *n*-butane and isobutane (C_4H_{10}), and pentanes (C_5H_{12}). The C_3, C_4, and C_5 hydrocarbons are removed before the gas is sold. The commercial natural gas supplied to the customer is therefore primarily a mixture of methane and ethane. The propane and butanes removed from natural gas are usually liquefied under pressure and sold as LPGs. Natural gas is found to consist mainly of the lower paraffins, with varying quantities of carbon dioxide, carbon monoxide, hydrogen, nitrogen, and oxygen, and in some cases also hydrogen sulfide and possibly ammonia. The chemical composition of natural gas is given in Table 2.2.

For hundreds of years, natural gas has been known as a very useful substance. The ancient peoples of Persia, Greece, and India discovered natural gas many centuries ago. About 2,500 years ago, the Chinese discovered that the energy in natural gas could be harnessed, and used to heat water. In the early days of the natural gas industry, the gas was mainly used to light streetlamps, and the occasional house. The Chinese piped the gas from shallow wells and burned it under large pans to evaporate seawater to obtain salt.

Natural gas was known in England as early as 1659, but it did not replace coal gas as an important source of energy in the world until after World War II.

Natural gas was first used in America in 1816 in gas lamps to illuminate the streets of Baltimore. Soon after, in 1821, William Hart dug the first successful American natural gas well in Fredonia, New York. The Fredonia Gas Light Company opened its doors in 1858 as the nation's first natural gas company. By 1900, natural gas had been discovered in 17 states. In the past 40 years, the use of natural gas has grown.

Historically, world natural gas reserves and natural gas usage have generally trended upward. Most of this usage of natural gas was near the wells that produced it owing to the lack of long-range pipelines or transport. In 1925, the first all-

Table 2.2 Chemical composition of natural gas

Component	Typical analysis (vol%)	Range (vol%)
Methane	94.9	87.0–96.0
Ethane	2.5	1.8–5.1
Propane	0.2	0.1–1.5
Isobutane	0.03	0.01–0.3
n-Butane	0.03	0.01–0.3
Isopentane	0.01	Trace to 0.14
n-Pentane	0.01	Trace to 0.14
Hexane	0.01	Trace to 0.06
Nitrogen	1.6	1.3–5.6
Carbon dioxide	0.7	0.1–1.0
Oxygen	0.02	0.01–0.1
Hydrogen	Trace	Trace to 0.02

welded pipeline over 200 miles in length was built, running from Louisiana to Texas. The growth in such pipeline networks and the cheap price of natural gas led to an expansion in its use. Between the turn of the twentieth century and 1970, usage and production of natural gas increased 50-fold. However, production of natural gas in the USA peaked in 1973, and by 1980, the USA began to import natural gas from other countries. Despite high rates of increase in natural gas consumption, particularly over the past decade, most regional reserves-to-production ratios are substantial.

2.3 Origin of Natural Gas

Natural gas is generally considered a nonrenewable fossil fuel. It comes from the decomposition of organic matter, just like oil and coal. Unlike oil and coal, though, it can come from almost any organic matter, whereas coal comes only from plant matter and oil comes almost exclusively from plankton and microplankton remains. Natural gas can come from both of these sources as well.

Decay and millions of years of geological stresses have transformed the complicated organic compounds that once made up living plants or animals into a mixture of alkanes. Natural gas is considered a fossil fuel because most scientists believe that natural gas was formed from the remains of tiny sea animals and plants that died 200–400 million years ago. When these tiny sea animals and plants died, they sank to the bottom of the oceans, where they were buried by layers of sediment that turned into rock. Over the years, the layers of sedimentary rock became thousands of feet thick, subjecting the energy-rich plant and animal remains to enormous pressure.

Most scientists believe that the pressure, combined with the heat of the earth, changed this organic mixture into petroleum and natural gas. Eventually, concentrations of natural gas became trapped in the rock layers like wet sponge traps water.

Natural gas can also come from unconventional sources. It is produced by decay of dead plant matter in swamps and rice fields. Animals, such as cattle and termites, produce large quantities as a by-product of digestion. These sources, though, cannot be tapped for energy use. Other unconventional sources, such as landfills, manure digesters, and wastewater treatment plants, are used to produce natural gas. China produces enough gas from manure digesters to provide cooking and lighting for over six million homes.

2.4 Natural Gas Resources

Natural gas is commercially produced from oil fields and natural gas fields. Gas produced from oil wells is called casinghead gas or associated gas. The natural gas industry is producing gas from increasingly more challenging resource types: sour gas, tight gas, shale gas, coal-bed methane, and methane gas hydrate. The world's largest gas field by far is Qatar's offshore North Field, estimated to have 25 trillion cubic meters of gas – enough to last more than 200 years at optimum production levels. The second-largest natural gas field is the South Pars Gas Field in Iranian waters in the Persian Gulf. It has estimated reserves of 8–14 trillion cubic meters of gas.

Natural gas is found around the world, but the largest reserves are in the former Soviet Union and the Middle East. Natural gas reserves are given Table 2.3. Since the mid-1970s, world natural gas reserves have generally trended upward each year. Natural gas reserves in the industrialized countries also increased between 2002 and 2003 by 18 trillion cubic feet (TCF). The decrease was largely offset by the enormous upward revision of gas reserves in Azerbaijan from 4 TCF in 2002 to 30 TCF in 2003 (IEO-AEO 2002).

Table 2.3 World natural gas reserves ($\times 10^6$ m^3) (IEA 2003)

Country	Reserve	Country	Reserve
Russia	47,573	Pakistan	71
Iran	23,002	India	65
Qatar	14,400	Former Yugoslavia	48
Saudi Arabia	6,216	Yemen	48
United Arab Emirates	6,000	Brunei	39
USA	5,196	Hungary	37
Algeria	4,500	Thailand	36
Venezuela	4,180	Papua New Guinea	35
Nigeria	3,500	Croatia	34
Iraq	3,100	Bangladesh	30
Turkmenistan	2,860	Myanmar	28
Australia	2,548	Austria	25
Uzbekistan	1,875	Syria	24
Kazakhstan	1,841	Ireland	20

Table 2.3 (continued)

Country	Reserve	Country	Reserve
The Netherlands	1,770	Vietnam	19
Canada	1,691	Slovakia	14
Kuwait	1,690	Mozambique	13
Norway	1,246	France	11
Ukraine	1,121	Cameroon	11
Mexico	835	The Philippines	10
Oman	821	Afghanistan	10
Argentina	777	Turkey	9
UK	736	Congo	9
Bolivia	680	Sudan	9
Trinidad and Tobago	665	Tunisia	8
Germany	343	Taiwan	8
Indonesia	262	Namibia	6
Peru	246	Rwanda	6
Italy	229	New Zealand	6
Brazil	221	Bulgaria	6
Malaysia	212	Israel	4
Poland	144	Angola	4
China	137	Equatorial Guinea	4
Libya	131	Japan	4
Azerbaijan	125	Côte d'Ivoire	3
Colombia	122	Ethiopia	3
Ecuador	105	Gabon	3
Romania	102	Ghana	3
Egypt	100	Czech Republic	3
Chile	99	Guatemala	3
Bahrain	91	Albania	3
Denmark	76	Tanzania	2
Cuba	71		

Another potentially large source of natural gas is methane hydrates. These are solid, crystalline features that are composed of a combination of methane, water ice, and other gases. The methane and gases are trapped in the lattice structure of the water ice, which, like coal beds, can hold much more natural gas than normal rock features. These hydrate formations are usually found on the ocean floor, where there is high pressure and near-freezing temperatures.

2.5 Natural Gas Production, Consumption, and Distribution

As is the case with oil, natural gas is unevenly distributed throughout the world. More than one third of the world's original gas endowment was in the territory of the former Soviet Union. The second-largest gas resource, located in the Middle East, comprised about 22% of the world total. About 38% of the world's remaining gas is in the former Soviet Union and 25% is located in the Middle East. South

America, Europe, Africa, and Asia/Oceania are each projected to contain less than 10% of the world's remaining gas.

The largest supplier of natural gas imports to western Europe is Russia, and those imports are continuing to grow. In the first 7 months of 2002, western European imports of Russian gas increased by 5.4% compared with the same period in 2001. Russia has plans to increase its presence in western European markets by building a pipeline that would bypass Ukraine and Poland (to avoid high transport fees and unauthorized diversion of gas) and initially transport gas from the Yamal Peninsula in western Siberia to Finland, Sweden, and Denmark. The intention is to extend the pipeline subsequently to the Netherlands via Germany and then along the floor of the North Sea to the UK (IEO 2003).

The world distribution of natural gas mirrors that of oil, which might be expected since oil and gas are often generated and reservoired together. However, the Middle East, although containing a very significant amount of gas, does not dominate world gas supply as it does world oil supply. The Worldwide Statistical Survey (WSS 2006) statistical survey contains the latest data on natural gas reserves, gross and marketed production, international trade, consumption, and prices of the main contracts in 1999. The evolution of gas transportation and distribution networks, underground gas storage capacities, as well as demand prospects are also compiled in this document. The study includes natural gas statistics for 102 countries and detailed comments for the major countries. World natural gas production has been maintained at a level of about 2,300 billion cubic meters a year during the last 5 years, and it is not expected to be affected by the decrease in production in Russia.

Beyond 2000 there are projections to 2100 by different authors, with modeling. These scenarios imply that the amount of gas produced from 2000 to 2300 would be 20 times the most likely technical estimate of the conventional and unconventional gas combined, making them utterly implausible. The present world gas production is not showing any drastic change.

Consumption of natural gas worldwide is projected to increase by an average of 2.8% annually from 2001 to 2025, compared with projected annual growth rates of 1.8% for oil consumption and 1.5% for coal consumption. Natural gas consumption in 2025, at 176 TCF, is projected to be nearly double the 2001 total of 90 TCF. The natural gas share of total energy consumption is projected to increase from 23% in 2001 to 28% in 2025.

Around the world, natural gas use is increasing for a variety of reasons, including prices, environmental concerns, fuel diversification and/or energy security issues, market deregulation, and overall economic growth. Figure 2.1 shows production and consumption trends of natural gas in the last few decades.

The USA consumes about 2.4 TCF more natural gas per year than it produces. Germany imports even more gas than the USA (2.6 TCF per year) and Japan slightly less (2.3 TCF per year). North America is the leading consumer of natural gas, but is also a leading producer. The former USSR region leads the world in gas production, and is second in consumption. Europe ranks third in natural gas consumption, but has to import 4.1 TCF per year. Asia/Oceania must also import

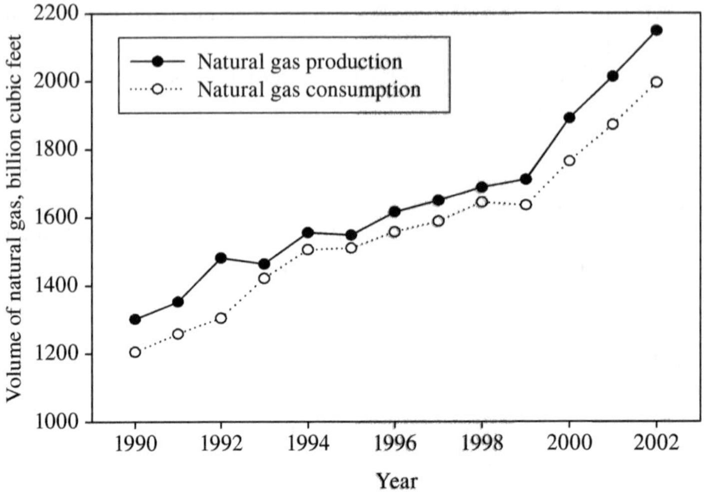

Fig. 2.1 Production and consumption trends of natural gas in the world

natural gas to satisfy demand. The other regions are relatively minor producers and consumers of gas. Compared with oil, only moderate amounts of natural gas are traded on world markets. The low density of gas makes it more expensive to transport than oil. A section of pipe in oil service can hold 15 times more energy than when it is used to transport high-pressure gas. For this reason, gas pipelines must have a larger diameter for a given energy movement. Compression adds to the disparity between the transportation costs of the two fuels. An oil pumping station uses energy to overcome frictional losses, but a gas line requires a large amount of energy to compress the gas before pipeline friction is even encountered.

Pipeline transportation is not always feasible because of the growing geographic distance between gas reserves and markets. Also, since potential political instabilities may affect long pipeline routes, importing countries may wish to diversify supply sources. Although natural gas can be piped in a gaseous state, it needs to be liquefied so that it can be economically transported by ship. A full LNG chain consists of a liquefaction plant, low-temperature, pressurized, transport ships, and a regasification terminal. World LNG trade is currently about 60 million metric tons per year, some 65% of which is imported by Japan. Other importers include France, Spain, Korea, Belgium, Taiwan, and Italy. Indonesia accounts for 39% of LNG exports, with Algeria in second place with 24%. Other exporters include Malaysia, Brunei, Australia, Abu Dhabi, and Libya. The USA imports and exports about one million metric tons of LNG per year. No grassroots LNG project has been commissioned since 1989 owing to intense competition with other fuels, notably oil.

The global market for natural gas is much smaller than that for oil because gas transport is difficult and costly, owing to the relatively low energy content in relation to volume. Currently, only about 16% of global gas production is traded internationally, with less than 4% of the trade accounted for by LNG. In spite of the high cost of gas transportation and the remote location of some future supply regions, increasing international trade in natural gas is expected.

The generation of electricity is an important use of natural gas. However, the electricity generated from natural gas is generally more expensive than that generated using coal because of increased fuel costs. Natural gas can be used to generate electricity in a variety of ways. These include (1) conventional steam generation, similar to coal-fired power plants in which heating is used to generate steam, which in turns runs a turbine with an efficiency of 30–35%, (2) a centralized gas turbine, in which hot gases from natural gas combustion are used to turn the turbine, and (3) a combined cycle unit, in which both steam and hot combustion gases are used to turn the turbine with an efficiency of 50–60%.

2.6 Compressed Natural Gas

CNG is odorless, colorless, and tasteless. It consists mostly of methane and is drawn from gas wells or is obtained in conjunction with crude oil production. An odorant is normally added to CNG for safety reasons.

Typically, to use CNG as a transportation fuel, natural gas is taken from a gas distribution system at pressures ranging from 0.3 to 1 MPa, compressed at a fueling station, and stored in cylinders on vehicles at pressures of about 20 MPa. Currently, the international NGV industry has not set a standardized CNG pressure. At 20 MPa, the volumetric energy density of CNG is about one fourth that of gasoline. Thus, with all efficiencies being equal, a CNG vehicle requires a tank 4 times the size of a gasoline tank for the same driving range. Liquefying natural gas by cooling it to approximately 400 K increases its energy density. LNG is typically stored at low pressures between 0.07 and 0.34 MPa to maintain its liquid state, but LNG is not available in most markets. Increasing the travel distance between refueling increases the investment, the weight of the storage tank, and the space needed to make it possible to use CNG. Natural gas has an octane value of 130, which is considerably higher than that of gasoline, which usually ranges between 84 and 97. The higher octane value of CNG generally gives very good engine performance characteristics (Demirbas 2002).

Despite the large difference in volumetric energy density between gasoline and CNG, the impact of CNG energy density on engine performance is less dramatic. As a gas, it has few cold-start problems. Its higher octane value allows for higher engine compression ratios than can be used with gasoline alone. Higher compression ratios allow for higher power and fuel efficiency. However, for the same compression ratio, the amount of natural gas air/fuel mixture that can be burned in each piston stroke is 10–15% less than for gasoline. Thus, there is a 10–15% loss of engine output power (Demirbas 2002).

To use CNG safely, technicians and drivers need to know what the differences are and how to work with them. Other subjects that must be understood are natural gas combustion and storage, working with high-pressure conduits, connectors, regulators, and cylinders, safety codes and industry standards, and recommended CNG cylinder inspections. Detailed hands-on training for installation and maintenance technicians should normally be provided by the conversion kit manufacturers (Demirbas 2002).

In general, the use of natural gas results in cleaner and longer-lasting engines. Less carbon builds up on spark plugs, in engine oil, and in the combustion chamber. Natural gas has higher ignition temperatures than gasoline, which increases the importance of maintaining proper ignition system operation.

The toxic emissions with CNG, without exception, are lower than for any other hydrocarbon fuel. This is a direct result of the fact that CNG is a single hydrocarbon, 90% methane, whereas all of the other fuels are a mix of hydrocarbons. LPG is a relatively simple mix of propane, butane, and pentane, in contrast to the complex mix that makes up the gasoline and diesel typically purchased at the service station. Combustion of gasoline and diesel results in emission into the air of methanol, formaldehyde, aldehydes, acrolein, benzene, toluene, xylene, etc., some of which are not yet part of any established emission standard but certainly are not beneficial to peoples' health (Demirbas 2002).

Soot emission from hydrocarbon flames is an important subject since it plays an important role in relation to both heat transfer by radiation and air pollution (Shahad and Mohammed 2000). The use of CNG in internal combustion engines permits operation with decreased NO_x emissions without increasing soot formation or specific fuel consumption.

The production, processing, transportation, and compression of natural gas to the CNG fuel that is used by vehicles results in less environmental impact than the production, transportation, and processing of crude oil and the transportation of gasoline or diesel to the service stations. Less carbon dioxide is produced by combustion of natural gas than by combustion of both diesel fuel and gasoline, which makes natural gas engines favorable also in terms of the greenhouse effect.

2.7 Liquefied Natural Gas

LNG is a product of natural gas which consists primarily of methane. Its properties are those of liquid methane, slightly modified by minor constituents. One property which differentiates LNG from LPG is the low critical temperature, about 200 K. This means that natural gas cannot be liquefied at ordinary temperatures simply by increasing the pressure, as is the case with LPG; instead, natural gas must be cooled to cryogenic temperatures to be liquefied and must be well insulated to be held in the liquid state. The reduction in volume makes natural gas much more cost-efficient to transport over long distances where pipelines do not exist. Where movement of natural gas by pipelines is not possible or economic, it can be transported by specially designed cryogenic sea vessels (LNG carriers) or cryogenic road tankers. LNG is shipped around the world in specially constructed seagoing vessels.

Until 2003, LNG prices closely followed oil prices. Since then, LNG prices to Europe and Japan have been lower than oil prices. In the USA in 2008, regular-grade motor gasoline retail prices averaged $ 3.26 per gallon, diesel fuel retail prices averaged $ 3.80 per gallon, and LNG retail prices averaged $ 3.36 per gallon (EIA 2009).

The natural gas fed into the LNG plant is treated to remove water, corrosive acid gases (H_2S and CO_2) and other components, such as dust, helium, and heavy hydrocarbons (ethane, propane, and butane). LNG typically contains more than 90% methane. The purification process can be designed to give almost 100% methane. The density of LNG is roughly 0.41–0.5 kg/L, depending on temperature, pressure, and composition. The heating value of LNG depends on the source of gas that is used and the process that is used to liquefy the gas. The higher heating value of LNG is 24 MJ/L at 109 K.

In its liquid state, LNG is not explosive and cannot burn. For LNG to burn, it must first vaporize, then mix with air in the proper proportions (the flammable range is 5–15%), and then be ignited.

Modern LNG storage tanks are typically the full containment type, which is a double-wall construction with a reinforced concrete outer wall and a high-nickel steel inner tank, with extremely efficient insulation between the walls. LNG must be kept cold to remain a liquid, independent of pressure. LNG is transported in specially designed ships with double hulls protecting the cargo systems from damage or leaks. Transportation and supply is an important aspect of the gas business, since LNG reserves are normally quite distant from consumer markets.

2.8 Contribution of Natural Gas to Global Warning

Table 2.2 shows the chemical composition of natural gas. Methane is by far the largest component, its presence accounting for about 95% of the total volume. Other components are ethane, propane, butane, pentane, nitrogen, carbon dioxide, water vapor, and traces of other gases. Very small amounts of sulfur compounds are also present. Since methane is the largest component of natural gas, the properties of methane are generally used when comparing the properties of natural gas with those of other fuels. Global warming has been increasingly associated with carbon dioxide levels. The gases (they consist of three or more atoms) with higher heat capacities than those of oxygen and nitrogen cause the greenhouse effect.

$$\begin{array}{llll} CH_4 & + 2O_2 \rightarrow & CO_2 & + 2H_2O \\ 1.00\,g & & 2.75\,g & \end{array} \tag{2.1}$$

$$\begin{array}{llll} 2C_4H_{10} & + 13O_2 \rightarrow & 8CO_2 & + 10H_2O \\ 1.00\,g & & 3.03\,g & \end{array} \tag{2.2}$$

$$\begin{array}{lll} C & + O_2 \rightarrow & CO_2 \\ 1.00\,g & & 3.66\,g \end{array} \tag{2.3}$$

From Eq. 2.1, among the fossil fuels, natural gas is the least responsible for CO_2 emissions. Natural gas is the cleanest-burning petroleum-based fuel. Combustion of LPG produces more CO_2 than combustion of natural gas (Eq. 2.2). The highest amount of CO_2 occurs according to Eq. 2.3. Thus, the responsibility of the fossil fuel increases with increasing carbon number. In addition, overall CO_2 emissions can be reduced by combustion of biomass because it is a CO_2-neutral fuel.

The percentage of carbon in LNG is the lowest in carbon-intensive fossil fuels (Davies 2001) and thus LNG has the lowest pollution potential when compared with other fossil fuels. Exhaust emissions from NGVs are much lower than those from equivalent gasoline-powered vehicles. For instance, NGV emissions of carbon monoxide are approximately 70% lower, non-methane organic gas emissions are 89% lower, and oxides of nitrogen emissions are 87% lower. In addition to these reductions in pollutants, NGVs also emit significantly lower amounts of greenhouse gases and toxins than do gasoline-powered vehicle. Dedicated NGVs produce little or no evaporative emissions during fueling and use. For gasoline-powered vehicles, evaporative and fueling emissions account for at least 50% of a vehicle's total hydrocarbon emissions. Dedicated NGVs also can reduce carbon dioxide exhaust emissions by almost 20%. Diesel exhaust is under review as a hazardous air pollutant.

Environmental concerns such as global warming have resulted in calls for increased use of natural gas. This is because when natural gas is burned, it gives off only half as much carbon dioxide per unit of energy produced as does coal, and 25% less than oil, virtually no sulfur dioxide, and only small amounts of nitrous oxides. Carbon dioxide is a greenhouse gas, whereas sulfur dioxide and nitrous oxides produced by oil and coal combustion cause acid rain.

Natural gas is mostly composed of methane and other light hydrocarbons. Both the carbon and the hydrogen in methane combine with oxygen when natural gas is burned, giving off heat. As mentioned earlier, reduced carbon dioxide emissions may not be the end of the story when it comes to natural gas and global warming. Methane is itself a greenhouse gas, and molecule for molecule can trap more heat than carbon dioxide and in this way counterbalances the carbon dioxide benefits of burning natural gas instead of oil and coal (Surmen and Demirbas 2003; Demirbas 2000).

2.9 Use of Natural Gas

For hundreds of years, natural gas has been known as a very useful substance. The Chinese discovered a very long time ago that the energy in natural gas could be harnessed, and used to heat water. In the early days of the natural gas industry, the gas was mainly used to light streetlamps, and the occasional house. However, with much improved distribution channels and technological advancements, natural gas is being used in ways never thought possible.

Table 2.4 Natural gas use in Europe by sector

Sector	Use (%)	Percentage of total
Industrial	43	38
Residential	22	25
Electricity generation	18	20
Commercial	14	15
Transportation	3	2

Natural gas is one of the most widely used forms of energy today. It is commonly used to heat and cool homes and businesses. Natural gas is a very versatile fuel which can be used for space and water heating, which have traditionally been the predominant uses of natural gas, representing about 75 and 15% of natural gas use, respectively. In the future, increasing concerns about urban air pollution may lead to increased use of natural gas by industry and for electricity generation, cooking, mechanical power, and heating.

The generation of electricity is the other main use of natural gas. Producing electricity from natural gas is generally more expensive than using other methods because of increased fuel costs. Natural gas can be used to generate electricity in many different ways. Natural gas power plants generate more than a couple of hundred megawatts using the same technology as coal-fired power plants. Natural gas is burned to produce heat, which boils water, creating steam, which passes through a turbine to generate electricity. Natural gas is one of the most widely used forms of energy today.

Natural gas is used by industrial, residential, electricity generation, commercial, and transportation sectors. Natural gas is used across all sectors, in varying amounts. Table 2.4 shows natural gas use by sector. The industrial sector accounts for the greatest proportion of natural gas use in the USA, with the residential sector consuming the second-greatest quantity of natural gas. The proportions of petroleum products, natural gas, coal nuclear power, and other energy sources consumed in the USA in 2002 were 39, 24, 22, 8, and 7%, respectively (EIA–AEO 2002).

2.9.1 Industrial Uses

Natural gas has innumerable uses in industry, and new applications are being developed every day. It is used to provide the base ingredients for such varied products as plastics, fertilizers, antifreeze, and fabrics. In fact, industry is the largest consumer of natural gas, accounting for 43% of natural gas use across all sectors. Natural gas is the second most used energy source in industry, trailing only electricity. Lighting is the main use of energy in the industrial sector, and accounts for the tremendous electricity requirements of this sector.

Industrial applications for natural gas also include the same uses as found in residential and commercial settings – heating, cooling, and cooking. Natural gas is also used for waste treatment and incineration, preheating of metals (particularly for iron and steel), drying and dehumidification, glass melting, food processing, and fueling industrial boilers. Natural gas may also be used as a feedstock for the manufacturing of a number of chemicals and products. Gases such as butane, ethane, and propane may be extracted from natural gas to be used as a feedstock for fertilizers and pharmaceutical products.

Natural gas as a feedstock is commonly found as a building block for methanol, which in turn has many industrial applications. Natural gas is converted to what is known as synthesis gas, which is a mixture of hydrogen and carbon oxides formed through a process known as steam reforming. In this process, natural gas is exposed to a catalyst, which causes oxidization of the natural gas when it is brought into contact with steam. This synthesis gas, once formed, may be used to produce methanol, which in turn is used to produce substances such as formaldehyde, acetic acid, and methyl *tert*-butyl ether, which is used as an additive for cleaner-burning gasoline. Methanol may also be used as a fuel source in fuel cells.

Industrial consumers reap great benefits from operating natural gas cogeneration or combined heat and power and combined cooling, heat, and power systems, similar to those used in commercial settings. For instance, natural gas may be used to generate electricity needed in a particular industrial setting. The excess heat and steam produced from this process can be harnessed to fulfill other industrial applications, including space heating, water heating, and powering industrial boilers.

Natural gas co-firing technologies are also helping to increase industrial energy efficiency, and reduce harmful atmospheric emissions. Co-firing is the process in which natural gas is used as a supplemental fuel in the combustion of other fuels, such as coal, wood, and biomass. For example, a traditional industrial wood boiler would simply burn wood to generate energy. However, in this type of boiler, a significant amount of energy is lost, and harmful emissions are very high. Adding natural gas to the combustion mix can have a twofold effect. Combustion of natural gas results in emission of fewer harmful substances into the air than combustion of a fuel such as wood. Since the energy needed to power the natural gas boiler remains constant, adding natural gas to the combustion mix can reduce harmful emissions.

2.9.2 Residential Use

Natural gas is one of the cheapest forms of energy available to the residential consumer. Table 2.5 shows the residential energy costs per unit. According to the US Department of Energy, the cost of natural gas costs is less than 30% of the cost of electricity. Not only is natural gas cheap for the residential consumer, it also has a number of varied uses. The best known uses for natural gas around the home are natural gas heating and cooking. Cooking with a natural gas range or oven can provide many benefits, including easy temperature control, self-ignition and self-cleaning, as well as costing approximately half as much as cooking with an electric range.

Table 2.5 Residential energy costs per unit

Electricity	Propane	Kerosene	No. 2 heating oil	Natural gas
100	39.3	37.5	32.1	27.1

Natural gas appliances are also rising in popularity owing to their efficiency and cost-effectiveness. Some examples of other natural gas appliances include space heaters, clothes dryers, pool heaters, fireplaces, barbecues, garage heaters, and outdoor lights.

Natural gas fuel cells and microturbines both offer residential consumers the possibility to disconnect from their local electricity distributor, and generate just enough electricity to meet their needs. Although this technology is still in its infancy, it is very promising in being able to offer independent, reliable, efficient, environmentally friendly electricity for residential needs.

2.9.3 Electricity Generation using Natural Gas

Natural gas can be used to generate electricity in a variety of ways. The most basic natural-gas-fired electricity generation consists of a steam generation unit, where fossil fuels are burned in a boiler to heat water and produce steam, which then turns a turbine to generate electricity. Natural gas may be used for this process, although these basic steam units are more typical of large coal or nuclear generation facilities. These basic steam generation units have fairly low energy efficiency. Typically, only 33–35% of the thermal energy used to generate the steam is converted into electric energy in these types of units.

Gas turbines and combustion engines are also used to generate electricity. In these types of units, instead of heating steam to turn a turbine, hot gases from burning fossil fuels (particularly natural gas) are used to turn the turbine and generate electricity.

Many of the new natural-gas-fired power plants are what are known as "combined-cycle" units. In these types of generating facilities, there is both a gas turbine and a steam unit, all in one.

Natural gas is one of the leading energy sources for distributed generation. Fuel cells are becoming an increasingly important technology for the generation of electricity. They are much like rechargeable batteries, except instead of using an electric recharger, they use a fuel, such as natural gas, to generate electric power even when they are in use. Fuel cells for distributed generation offer a multitude of benefits, and are an exciting area of innovation and research for distributed generation applications.

The industrial sector, which is the world's largest consumers of natural gas, will account for 43% of projected natural gas use in 2030. In the electric power sector, natural gas is an attractive choice for new generating plants because of its relative fuel efficiency and low carbon dioxide emissions. Electricity generation will account for 35% of the world's total natural gas consumption in 2030 (Demirbas 2006).

2.9.4 Commercial Use

Commercial uses of natural gas are very similar to residential uses. The commercial sector includes public and private enterprises, including office buildings, schools, churches, hotels, restaurants, and government buildings. The main uses of natural gas in this sector include space heating, water heating, and cooling. For restaurants and other establishments that require cooking facilities, natural gas is a popular choice to fulfill these needs. Natural gas is the primary energy source for space and water heating, cooking, and drying, and also accounts for about 13% of energy used in commercial cooling.

Space heating accounts for 36% of commercial energy use, lighting for 19%, cooling for 12%, water heating for 8%, cooking for 6%, drying for 3%, and other uses for 16%.

2.9.5 Natural Gas in the Transportation Sector

Natural gas has long been considered an alternative fuel for the transportation sector. Most NGVs operate using CNG. This compressed gas is stored in a tube-shaped storage tank, in a similar fashion to a car's gasoline tank, attached to the rear, top, or undercarriage of the vehicle. A CNG tank can be filled in a similar manner, and in a similar amount of time, as a gasoline tank.

NGVs offer many benefits, from improving public health and the environment to aiding the transition to fuel cell vehicles. Compared with vehicles fueled by conventional diesel and gasoline, NGVs can produce significantly lower amounts of harmful emissions such as nitrogen oxides, particulate matter, and toxic and carcinogenic pollutants. NGVs can also reduce emissions of carbon dioxide, the primary greenhouse gas. NGV and natural gas infrastructure development can facilitate the transition to this technology.

Despite the large difference in volumetric energy density between gasoline and CNG, the impact of CNG energy density on engine performance is less dramatic. As a gas, it has few cold-start problems. Its higher octane value allows for higher engine compression ratios than can be used with gasoline alone. Higher compression ratios allow for higher power and fuel efficiency. However, for the same compression ratio, the amount of natural gas air/fuel mixture that can be burned in each piston stroke is 10–15% less than for gasoline. Because of this, there is a 10–15% loss of engine output power.

2.10 Importance of Natural Gas

Concerns about acid rain and global warming will no doubt result in increased use of natural gas in the future. Two areas which could see expanded use of natural gas are fuel cells and transportation. Fuel cells are used to generate electricity, and operate something like a battery. The difference is that the energy for fuel cells comes from hydrogen, which can be made from natural gas. Fuel cells eliminate the need for turbines and generators, and can operate at efficiencies as high as 60%. Fuel cells also operate at low temperatures, thus reducing emissions of acid rain causing nitrous oxides, which are formed during high-temperature combustion of any fuel.

Concern over urban air pollution may lead to increased use of natural gas as a transportation fuel in the future. Natural gas burns far more cleanly than gasoline and diesel fuel, producing fewer nitrous oxides, unburned hydrocarbons, and particulates. NGVs require large storage tanks for their fuel. Therefore, the main market may not be for private use, but may be for vehicles and trucks which are used within cities.

Advances in technology over the last two decades and changes in the energy industry are leading to natural gas becoming the best cooling material. Natural gas cooling equipment is available in sizes to meet virtually every need – from air conditioning in residential homes to large-scale industrial refrigeration and process cooling. Today's natural gas cooling equipment is efficient and economical. Natural gas systems can save as much as 50% over the cost of conventional electric cooling equipment.

2.11 Environmental Impacts

Natural gas is the cleanest-burning alternative fuel. Exhaust emissions from NGVs are much lower than those from equivalent gasoline-powered vehicles. For instance, NGV emissions of carbon monoxide are approximately 70% lower, non-methane organic gas emissions are 89% lower, and oxides of nitrogen emissions are 87% lower. In addition to these reductions in pollutants, NGVs also emit significantly lower amounts of greenhouse gases and toxins than do gasoline vehicles. Dedicated NGVs produce little or no evaporative emissions during fueling and use. For gasoline vehicles, evaporative and fueling emissions account for at least 50% of a vehicle's total hydrocarbon emissions. Dedicated NGVs can also reduce carbon dioxide exhaust emissions by almost 20%. Diesel exhaust is under review as a hazardous air pollutant.

Per unit of energy, natural gas contains less carbon than any other fossil fuel, and thus produces lower carbon dioxide emissions per vehicle mile traveled. Although NGVs do emit methane, another principal greenhouse gas, any slight increase in methane emissions would be more than offset by a substantial reduc-

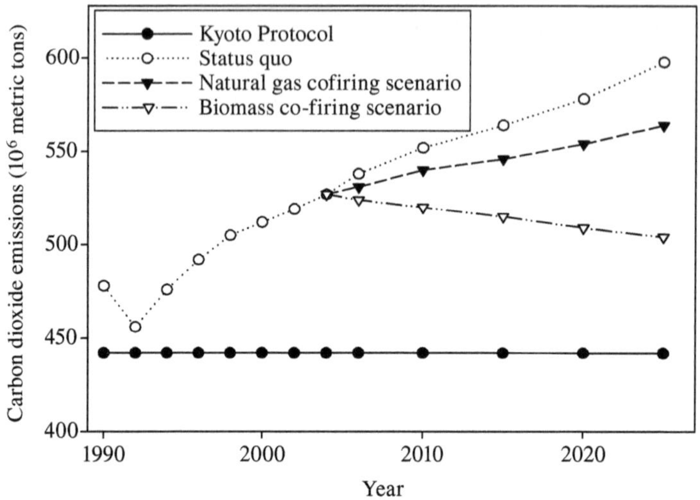

Fig. 2.2 Scenarios of co-firing of natural gas and biomass as fuels to reduce fossil-fuel-based CO_2 emissions

tion in carbon dioxide emissions. NGVs also emit very low levels of carbon monoxide (approximately 70% lower than a comparable gasoline-powered vehicle) and volatile organic compounds. Although these two pollutants are not themselves greenhouse gases, they play an important role in helping to break down methane and some other greenhouse gases in the atmosphere, and thus increase the global rate of methane decomposition. Figure 2.2 shows plots of scenarios of co-firing of natural gas and biomass as fuels to reduce fossil-fuel-based carbon dioxide emissions.

Natural gas could be considered the most environmentally friendly fossil fuel, because it has the lowest carbon dioxide emissions per unit of energy and because it is suitable for use in high-efficiency combined-cycle power stations. Because of the energy required to liquefy and to transport LNG, its environmental performance is inferior to that of natural gas, although in most cases LNG is still superior to alternatives such as fuel oil or coal. CNG has a much lower environmental impact than other hydrocarbon fuels when the process from production in the fields to filling of the vehicle tanks is taken into account. The toxic emissions with CNG, without exception, are lower than for any other hydrocarbon fuel. This is a direct result of the fact that CNG is a single hydrocarbon, 90% methane, whereas all of the other fuels are a mix of hydrocarbons.

Soot emission from hydrocarbon flames is an important subject since it plays an important role in relation to both heat transfers by radiation and air pollution (Shabad and Mohammed 2000). The use of CNG in internal combustion engines permits operation with decreased NO_x emissions without increasing soot formation or specific fuel consumption. The production, processing, transportation, and compression of natural gas to the CNG fuel that is used by vehicles results in less environmental impact than the production, transportation, and processing of crude oil and the transportation of gasoline or diesel to the service stations. Less carbon dioxide is produced by combustion of natural gas than by combustion of both diesel fuel and gasoline, which makes natural gas engines favorable in terms of the greenhouse effect.

2.12 Summary

Natural gas is the fastest-growing primary energy source in the International Energy Outlook 2003 (IEO 2003) forecast. Natural gas consumption is projected to nearly double between 2001 and 2025, with the most robust growth in demand expected among the developing nations.

Natural gas is formed deep underground, usually in areas around coal and oil. It is composed primarily of methane, but also contains other chemical species, such as butane and propane. If the mixture is composed only of these species, it is called dry natural gas, as there will be no liquid components at standard pressure and temperature. Natural gas might also contain nonhydrocarbon compounds, such as water vapor, carbon dioxide, and hydrogen sulfide.

Over the next 20 years, the role of natural gas in global energy consumption will increase substantially. The speed of the transition to natural gas will be driven by environmental constraints, increased demand, and new technologies. A potential source of natural gas lies in the enormous worldwide gas hydrate reserves. It is estimated that the size of this resource ranges up to 20,000 trillion cubic meters. These deposits can cause problems and safety concerns relating to drilling, production of oil and gas, and building and operation of pipelines. Naturally occurring gas hydrates are normally found at the seafloor or in shallow sediments where the pressures and temperatures are conducive to hydrate formation.

Natural gas is generally considered a nonrenewable fossil fuel. Most scientists believe that natural gas was formed from the remains of tiny sea animals and plants that died 200–400 million years ago. When these tiny sea animals and plants died, they sank to the bottom of the oceans, where they were buried by layers of sediment that turned into rock. Over the years, the layers of sedimentary rock became thousands of feet thick, subjecting the energy-rich plant and animal remains to enormous pressure, and transforming their constituent compounds into a mixture of alkanes.

For hundreds of years, natural gas has been known as a very useful substance. The Chinese discovered a very long time ago that the energy in natural gas could be harnessed, and used to heat water. In the early days of the natural gas industry, the gas was mainly used to light streetlamps, and the occasional house. However, with much improved distribution channels and technological advancements, natural gas is being used in ways never thought possible.

Natural gas is used in industrial, residential, electricity generation, commercial, and transportation sectors. Natural gas is used across all sectors, in varying amounts. The industrial sector accounts for the greatest proportion of natural gas use in the USA, with the residential sector consuming the second-greatest quantity of natural gas.

Concerns about acid rain and global warming will no doubt result in increased use of natural gas in the future. Two areas which could see expanded use of natural gas are fuel cells and transportation. Fuel cells are used to generate electricity, and operate something like a battery. The difference is that the energy for fuel cells comes from hydrogen, which can be made from natural gas. Fuel cells eliminate the need for turbines and generators, and can operate at efficiencies as high as 60%.

References

Davies P (2001) The new challenge of natural gas. Paper presented at OPEC and the global energy balance: towards a sustainable future, Vienna, 28 September 2001

Demirbas A (2000) Recent advances in biomass conversion technologies. Energy Educ Sci Technol 6:19–40

Demirbas A (2002) Fuel properties of hydrogen, liquefied petroleum gas (LPG), and compressed natural gas (CNG) for transportation. Energy Sources 24:601–610

Demirbas A (2006) Global biofuel strategies. Energy Educ Sci Technol 17:27–63

EIA (2009) Official energy statistics from the US Government. Energy Information Administration, Washington

EIA–AEO (2002) Natural gas uses. Energy Information Administration–International Energy Outlook, Washington

IEA (2003) Gas statistics. International Energy Agency, Paris

IEO (2003) Natural gas trends. http://www.eia.doe/iea/overview.html. Accessed May 2003

Kvenvolden K (1993) Gas hydrates – geological perspective and global change. Rev Geophys 31:173–187

Shahad HAK, Mohammed YKA (2000) Investigation of soot formation and temperature field in laminar diffusion flames of LPG-air mixture. Energy Convers Manage 41:1897–1916

Surmen Y, Demirbas A (2003) Cofiring of biomass and lignite blends: resource facilities; technological, environmental and economic issues. Energy Sources 25:175–187

WSS (2006) Natural gas in the world. http://www.cedigaz.org/about_us/about_us.htm

Chapter 3
Gas Hydrates

3.1 Introduction

The most important alternative energy resources for the near future are renewable energy (especially biomass, algae, wind, and photovoltaics), hydrogen, and methane hydrates.

Renewable energy is a promising alternative solution because it is clean and environmentally safe. It also produces lower or negligible levels of greenhouse gases and other pollutants when compared with the fossil energy sources it is replacing. Approximately half of the global energy supply will be from renewables in 2040 according to the European Renewable Energy Council (2006). The most significant developments in renewable energy production will be observed in photovoltaics (from 0.2 to 784 million tons oil equivalent) and wind energy (from 4.7 to 688 million tons oil equivalent) between 2001 and 2040. The majority of biomass energy is produced from wood and wood waste (64%), followed by municipal solid waste (24%), agricultural waste (5%), and landfill gases (5%). It is increasingly understood that first-generation biofuels (produced primarily from food crops such as grains, and oil seeds) are limited in their ability to achieve targets for biofuel production, climate change mitigation, and economic growth. Their sustainable production is under review and is being hotly debated globally. The cumulative impacts of these concerns have increased interest in developing next-generation biofuels produced from systematically grown energy crop nonfood feedstocks such as algae and *Jatropha* species for biodiesel. Researchers are also focused on converting agricultural residues and municipal and industrial waste to biofuels. These "next-generation biofuels" could avoid many of the concerns facing first-generation biofuels and potentially offer the greatest opportunities in the longer term.

Hydrogen as a potential energy source, providing a transformation from an oil to a hydrogen economy, has been a topic under consideration for more than 25 years. Hydrogen as a fuel provides clean energy without greenhouse gas emissions. Although hydrogen is abundantly present on the earth's surface as water, cost-effective extraction of hydrogen for transportation and industrial use presents major technological challenges. The advantages of hydrogen power are the high energy content of hydrogen per unit mass, the absence of pollution during energy conversion, the excellent flammability, flame speed, and autoignition temperature, and the potential use of hydrogen in fuel cells. However, hydrogen has the lowest energy content per unit volume, which is critical for the transportation industry.

A new material (methane hydrates) is abundant on the earth, and, when extracted, promises to potentially solve the world's energy needs for many centuries. This material is abundantly present in the continental margins of the ocean and in the permafrost regions of the Arctic, packed with energy to meet future energy demands.

If it is economically extracted from the ocean sediments, the over 160 m^3 of methane trapped within each 1 m^3 of hydrate constitutes a large deposit of an alternative energy source. The US Geological Survey estimates the resource potential in the USA to be about 200,000 TCF. For comparison, the current annual consumption of natural gas is about 22 TCF. On the basis of these estimates, at about 1% recovery, the deposit has the potential to fill the natural gas needs of the USA at the present rate of consumption for the next 100 years. Additionally, direct fuel combustion of methane not only provides high energy density per weight, but also contributes minimum emission of carbon dioxide as a by-product; nearly 30 times less than gasoline and 60 times less than coal.

Currently, natural gas is the fastest-growing primary energy source in the world. Natural gas is widely available and it is the cleanest-burning petroleum-based fuel. Natural gas can be produced from natural gas hydrates, or methane hydrates. The presence of gas hydrates in oceanic sediments was first postulated on the basis of seismic observations. Gas hydrates represent one of the world's largest untapped reservoirs of energy and, according to some estimates, have the potential to meet global energy needs for the next 1,000 years. Gas hydrates include combustible matter, and are potentially one of the most important energy resources for the future. Gas hydrates are of considerable interest because of their potential as an energy resource and because of their potential role in global climate change. From an energy resource point of view, the enormous amounts of gas hydrate under the ocean and beneath Arctic permafrost represent an estimated more than 50% of all carbonaceous fuel reserves on Earth.

Naturally occurring gas hydrates are a form of water ice which contains a large amount of methane within its crystal structure. Gas hydrates are crystalline solid compounds, consisting of a gas molecule surrounded by a cage of water molecules. They are stable under the high pressure and low temperature typical of deepwater sediments of the world ocean. Methane gas hydrate is composed of natural gas molecules trapped inside ice. Per unit mass, gas hydrates contain twice as much carbon as all other fossil fuels (coal, natural gas, and oil) combined. Table 3.1 shows worldwide amounts of organic carbon sources. The worldwide amount of organic carbon bound in gas hydrates is conservatively estimated to total twice the amount of carbon to be found in all known fossil fuels on Earth (Satoh *et al.* 1996; Kvenvolden 1998; Collett 2002).

Table 3.1 Worldwide amounts of organic carbon sources

Source of organic carbon	Amount (gigatons)
Gas hydrates (onshore and offshore)	10,000–11,000
Recoverable and nonrecoverable fossil fuels (oil, coal, gas)	5,000
Soil	1,400
Dissolved organic matter	980
Land biota	880
Peat	500
Others	70

The amount of methane trapped in marine sediments as a hydrate represents such an immense carbon reservoir that it must be considered a dominant factor in estimating unconventional gas energy resources. Significant safety and environmental concerns are also associated with the presence of natural gas hydrates, ranging from their possible impact on the safety of conventional drilling operations to their influence on the earth's periodic natural releases into the atmosphere of large volumes of hydrate-sourced methane or its derivative carbon dioxide. Methane gas hydrates are increasingly considered to be a potential energy resource (Hacisalihoglu *et al.*, 2008). New robot mining methods could be developed to obtain energy from methane hydrate.

Gas hydrates contain highly concentrated methane, which could be important both as an energy resource and as a factor in global climate change. Because gas hydrates change the stiffness of sediments, they may also be important in understanding hazards to seafloor installations in deep water such as wells, pipelines, and drilling platforms. The gas hydrate deposits potentially offer a vast new source of low-polluting, carbon-based energy that could provide a comfortable future. General interest is currently focusing on gas and gas hydrates contained in marine sediments. This is mainly due to (1) the environmental consequences of gas release at the seafloor, as a contributor to the greenhouse effect, (2) their impact on the seafloor stability, as a potential cause of submarine geohazards, and (3) the fuel resource potential of the gas hydrates, as they contain a great volume of methane.

3.1.1 Definition

Gas hydrates are icelike crystalline solids that form from mixtures of water and light natural gases such as methane, carbon dioxide, ethane, propane, and butane. Gas hydrates occur where pressure, temperature, gas saturation, and local chemical conditions combine to make them stable. Gas hydrates occur in the pore spaces of sediments, and may form cements, nodes, veins, or layers. Methane is the dominant component among other hydrocarbon gases in sediments (Mazzini *et al.* 2004; Popescu *et al.* 2006).

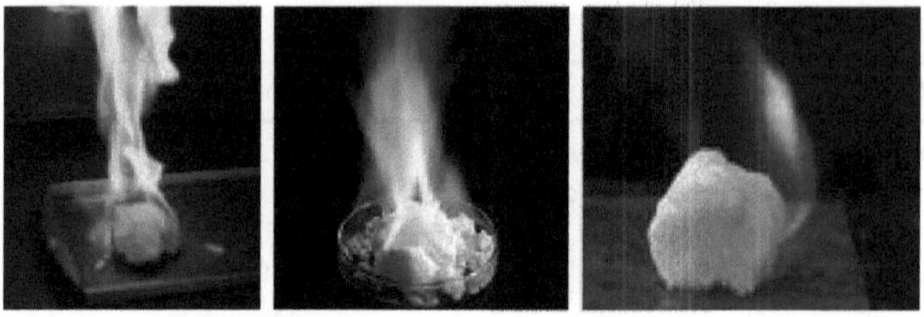

Fig. 3.1 Methane gas hydrate samples

Given the favorable combination of pressure and temperature, and the availability of free methane and water, gas hydrates can form and remain stable (Sloan 1990). Such conditions can exist in ocean-bottom sediments at water depths below 500 m (Kvenvolden 1993a). The terms "methane hydrate" and "gas hydrate" are often used interchangeably, and refer to the methane–water crystalline structure called a clathrate.

Gas hydrates are stable only under specific pressure–temperature conditions. Under the appropriate pressure, they can exist at temperatures significantly above the freezing point of water. The maximum temperature at which a gas hydrate can exist depends on the pressure and the gas composition. For example, methane plus water at 41 atm forms a hydrate at 278 K, whereas at the same pressure, methane plus 1% propane forms a gas hydrate at 281 K. Hydrate stability can also be influence by other factors, such as salinity (Edmonds *et al.* 1996). Per unit volume, gas hydrates contain a tremendous amount of gas. For example, 1 m^3 of hydrate disassociates at atmospheric temperature and pressure to form 164 m^3 of natural gas plus 0.8 m^3 of water (Kvenvolden 1993a). With pressurization, gas hydrates remain stable at temperatures up to 291 K.

As can be seen from Fig. 3.1, methane gas hydrate samples easily burn under barometric pressure. If methane hydrate is either warmed or depressurized, it will revert back to water and natural gas. Although global estimates vary considerably, the energy content of methane occurring in hydrate form is immense, possibly exceeding the combined energy content of all other known fossil fuels.

3.1.2 History

Study of gas hydrates started in 1778 (by Priestly), whereas the chemistry of gas hydrates was discovered in 1810 by Sir Humphrey Davy, who found that icelike crystals were formed when an aqueous solution was cooled, and they were considered to be a laboratory curiosity.

Table 3.2 shows the milestones in the history of methane gas hydrates. The hydrates were first studied thoroughly in the 1890s. At that time they were considered a laboratory oddity, mostly because they often formed well above the freezing point of water. It was not until 1930, when natural gas pipelines were extended

Table 3.2 Milestones in the history of methane gas hydrates

Year	Progress
1888	Methane gas hydrate first discovered by Villard
1934	Hydrates discovered to plug gas pipelines
1952	sI crystal unit confirmed
1964	Hydrates found in Siberian permafrost
1970	Oceanographers drill through a methane gas hydrate deposit
1981	A drilling vessel in Guatemala recovers a methane hydrate sample intact
1992	Beginning of intentionally looking for methane gas hydrate deposits

sI structure I

into colder climates, that methane hydrates received more attention. Clathrate formation turned out to be a major problem, clogging pipelines during transportation of gas under cold conditions. Even with this discovery, the clathrates were considered a nuisance rather than a resource.

Clathrates are crystalline solids which look like ice, and which occur when water molecules form a cagelike structure around smaller "guest molecules." The most common guest molecules are methane, ethane, propane, isobutane, *n*-butane, nitrogen, carbon dioxide, and hydrogen sulfide, of which methane occurs most abundantly in natural hydrates (Mahajan *et al.* 2007).

In 1964, naturally occurring methane gas hydrate was discovered by a Russian drilling crew in Messoyakha, a Siberian gas field. Interest in this subject deepened and scientists set out to find other locations around the world where methane could be found in the frozen form.

The presence of gas hydrates in oceanic sediments was first postulated on the basis of seismic observations (Tucholke *et al.* 1977). It soon appeared that vast amounts of gas hydrates must occur in sediments of continental slopes and rises if the bottom-simulating reflector and the commonly underlying acoustic transparent layer were to be attributed to gas hydrates and underlying gas-charged sediments (Shipley *et al.* 1979). In the 1970s, methane hydrates were found in ocean sediments. In 1992, the Ocean Drilling Program began intentionally looking for hydrate deposits, and samples were brought to the surface for further study (Farkhondeh and Gheisi 2002). The intensive exploration activity undertaken by oceanographers and petroleum geologists during the last decade facilitated the discovery of widespread occurrence of gas hydrates in the continental margin areas (Gupta 2004).

3.1.3 Gas Hydrate Origin and Formation

Methane is formed from the microbial decay of organic matter in the absence of oxygen and can also be of thermogenic origin (Gupta 2004). The methane in gas hydrates is dominantly generated by bacterial anaerobic degradation of organic matter in low-oxygen environments. Bacterial gas formed during early diagenesis of organic matter can become part of a gas hydrate in continental shelf sediment. Similarly, thermogenic gas leaking to the surface from a deep thermogenic gas accumulation can form a gas hydrate in the same continental shelf sediment. There are hydrate forms in both primary and secondary pore spaces and fractures in sediments as a diagenetic mineral (Sloan 1998). Methane hydrate forms at high pressure and low temperature, where sufficient gas is present, and in general in two types of geological settings: in the Arctic, where hydrate forms beneath the permafrost, and beneath the ocean floor at water depths greater than about 500 m. The hydrate deposits themselves may be several hundred meters thick.

As already stated, the methane in gas hydrates is dominantly generated by bacterial degradation of organic matter in low-oxygen environments. It could be that the methane is produced by bacteria near the seafloor that are decomposing organic sediments. However, it might also be that the methane originates from oil deposits deep within Earth that leak to the seafloor bottom through faults and cracks. Either way, it is still unknown how the methane becomes trapped within an ice lattice. This lack of knowledge means that we have no idea of what conditions are favorable for the formation of methane gas hydrates. Organic matter in the uppermost few centimeters of sediments is first attacked by aerobic bacteria, generating carbon dioxide, which escapes from the sediments into the water column. In this region of aerobic bacterial activity, sulfates are reduced to sulfides. If the sedimentation rate is low, the organic carbon content is low (less than 1%), oxygen is abundant, and aerobic bacteria use up all the organic matter in the sediments. But where sedimentation rates and the organic carbon content are high, the pore waters in the sediments are anoxic at depths of only a few centimeters, and methane is produced by anaerobic bacteria. Below the zone of solid clathrates, large volumes of methane may occur as bubbles of free gas in the sediments (Matsumoto *et al.* 1996; Dickens *et al.* 1997).

Clathrates occur wherever the conditions within the sediments are in the methane clathrate stability field, and where methane and water are available. This stability is limited by temperature and pressure: gas hydrates are stable at low temperatures and/or high pressures. Because of the requirements of pressure and temperature, and because of the requirement of relatively large amounts of organic matter for bacterial methanogenesis, clathrates are mainly restricted to two regions: (1) high latitudes and (2) along the continental margins in the oceans (Kvenvolden 1998).

Methane is an important part of the natural carbon cycle, but much remains unknown about the processes leading to methane generation, its transport and transformations within sediments, and its significance in global carbon budgets (Hovland and Judd 1998; Dickens 2003; Torres *et al.* 2004). Organic-rich sediments at water depths greater than 500 m usually host methane as a clathrate (Sloan 1998).

Subseabed methane within the continental margin sediments is produced primarily by microbial or thermogenic processes. In the microbial process, organic debris is decomposed by a complex sequence (methanogenesis) into methane by bacteria in an anoxic environment. Organic matter is composed of carbon, hydrogen, and phosphorus in a ratio of 106:16:1, and decomposition results in production of methane (Desa 2001):

$$(CH_2O)_{106}(NH_3)_{16}(H_3PO_4) \rightarrow 53CO_2 + 53CH_4 + 16NH_3 + H_3PO_4. \quad (3.1)$$

Acetate fermentation, a stage in the decomposition process, also produces methane:

$$CH_3COOH \rightarrow CH_4 + CO_2. \quad (3.2)$$

Further reduction of carbon dioxide also produces methane:

$$CO_2 + 4H_2 \rightarrow CH_4 + 2H_2O. \quad (3.3)$$

In the thermogenic process, thermal cracking of organically derived materials forms petroleum hydrocarbons (including methane). This generally occurs at considerable depth (more than 2 km) in sedimentary basins where temperatures exceed 273 K. Thermogenic methane may also be derived by thermal degradation of oil at even greater depths, and by the maturation of coal (Desa 2001).

After generation, methane in sediment can be transported through the movement of pore-water-containing dissolved gas, free gas flow, and/or molecular diffusion. When ascending, methane molecules reach favorable conditions and hydrate will form within the pore spaces of sediments in the presence of water molecules. Thus, biogenic methane formation may take place both *in situ* and beneath the hydrate stability zone, whereas thermogenic methane must be formed below the hydrate stability zone, and move up into it. Geochemical and isotopic techniques are used to identify the biogenic or thermogenic origin of the methane. Hydrate progressively fills and cements the sediment pore spaces and fractures, giving rise to massive and vein-type hydrate deposits.

The temperature and pressure conditions for hydrate stability depend on the composition of the gas and on the presence of salts and other components in seawater. It is generally believed that pore water has to be fully saturated with methane before natural hydrate can form. This condition can be met by (1) supply of sufficiently large amounts of organic matter in the sediments to generate enhanced methanogenic decomposition or (2) large upward methane fluxes, mostly related to fault zones, or other conduits, such as diapirs and mud volcanism. Free methane in a geological formation exists within the pores of low-density rocks. Any hydrate layer may trap free methane as long as the layer forms a seal through which gas cannot migrate. The free methane may be thermogenic gas that has migrated upward from the earth's crust, or it may be biogenic gas that was previously a hydrate layer but has now melted (Desa 2001).

Gas hydrates occur in two discrete geological situations: (1) marine shelf sediments (Kvenvolden 1993b; Kvenvolden and Lorenson 2000) and (2) onshore polar regions beneath permafrost. Hydrates occur in these two types of settings because these are the settings where the pressure–temperature conditions are within the hydrate stability field (Lerche and Bagirov 1998). Gas hydrates can be detected seismically as well as using well logs (Goldberg and Saito 1998; Hornbach *et al.* 2003). In polar media, the gas hydrates are commonly linked to permafrost occurrence onshore and on the continental shelves. In the oceans, gas hydrates are found on the outer continental margins, where the supply of organic material is high enough to generate enough methane, and water temperatures are close to the freezing point.

Gas hydrate formation is usually described as a crystallization process with nucleation, growth, agglomeration, and breakage. Gas is dissolved in water, and nucleation starts primarily at the gas–water interface, where the gas concentration is highest. The recently formed crystals disperse in the bulk liquid and start to grow as more gas is supplied to the liquid phase. Usually, after a while, agglomeration and breakage of the growing particles can be observed. In the literature, the gas consumption rate and the particle size distribution of the hydrate crystals formed are used to describe the hydrate formation process quantitatively.

The formation rate of natural gas hydrate is governed by a multitude of factors, including the pressure, temperature and gas composition, also called PVT effects. Also, the rate of hydrate formation is determined by the combined effects of heat and mass transfer. Cooling is required to remove the hydrate heat of formation. Mass transport is required to dissolve the natural gas in liquid water, and to bring the dissolved gas molecules into contact with a growing hydrate crystal. In addition to the aforementioned factors, the rate of hydrate formation depends on the nature of crystal growth, also referred to as chemical reaction kinetics. The overall rate of hydrate formation, therefore, depends on PVT effects, transport effects, and reaction effects.

It was found that pressure, temperature driving force, and gas–liquid interfacial area were the most important parameters affecting the gas consumption rate. Increased stirring rate and pressure resulted in an increased rate of consumption of gas, whereas an increase in temperature resulted in a decreased consumption rate. The effect of stirring was due to the increased gas–liquid interfacial area with increased stirring rate.

The clathrate stability depends on temperature and pressure. The average thickness of the clathrate stability zone along continental margins is about 500 m. The depth of its lower boundary is determined by the geothermal gradient: at greater depths in the sediment, the temperature becomes too high for gas hydrates. The thickest clathrate zones occur in regions of low geothermal gradients; most clathrates occur within 2,000 m of the earth's surface.

Hydrate concentrates methane produced in the deep biosphere by bacterial decomposition of organic matter (Rostrup-Nielsen 1984; Peña *et al.* 1996), especially on passive continental margins where plunging plate boundaries and subduction zones are not present to provide pathways for deep-sourced thermogenic methane. Gaseous methane migrates upward, and is concentrated within and immediately beneath a zone in which hydrate is thermodynamically stable. This is the hydrate stability zone that extends downward from the cold seafloor at water depths greater than about 500 m in nonpolar, open ocean conditions, and in permafrost regions from some depth within the water-ice permafrost zone (about 200 m) to some depth that is determined locally by rising temperature. Permafrost hydrates exist at low pressures and temperatures. They occur as part of a compound water-ice and hydrate permafrost on land and on continental shelves of Alaska, Canada, and Russia. Methane hydrate and water ice form a compound cryogenic zone. Water ice is stable at the surface at about 273 K, whereas hydrate is stable from some depth below the surface to some depth below the base of the water-ice stability zone. In Alaskan permafrost, local thermal variations result in maximum hydrate stability depths of 600–1,075 m with associated crustal temperatures of approximately 285–287 K (Bradford and Vannice 1999).

Oceanic hydrate contains up to 95% of all naturally occurring hydrate worldwide. On cold planets such as Mars, however, where no deep oceans now occur, gas hydrate would be found entirely in the cryosphere and its analog on Earth is hydrate that is found in permafrost regions.

The response of gas hydrate formations to tectonic, gravitational, and anthropogenic forces is a function of the mechanical strength and geometric distribution of the phases (sediment, gas hydrate, water) present and the cohesion and frictional resistance between grains (Durham *et al.* 2003). The concentrations of distributed gas hydrate in formations identified as hydrate-bearing are poorly known because of the difficulty in making direct observations; recent estimates put the hydrate concentration as low as 1–2% by volume in marine sediments (Holbrook *et al.* 1996) and as high as several tens of percent in permafrost regions (Dallimore and Collett 1999). In deposits where hydrate is located structurally between sediment grains (Helgerud 2001) or as cementation around contacting grains (Guerin *et al.* 1999), grain-to-grain cohesion and frictional resistance, from which soft sediments derive their strength (Martinson 1994; Mulder and Alexander 2000), can be affected, implying an effect on formation strength even at low concentrations of intergranular hydrate. At higher concentrations, a formation-strengthening effect from a solid phase in pores can be expected by analogy to strength enhancement in frozen soils at volume concentrations above 20–30% (Andersland and Ladanyi 1994) and from general theory and experience with rheology of mixtures (Handy 1994; Tullis *et al.* 1991).

Methane hydrate was formed by increasing methane pressure in a vessel to 6.2 MPa while maintaining the temperature at 274.3 K (within the hydrate stability region). This method is similar to that used by Handa and Stupin (1992), and has been thought to produce hydrate that surrounds and cements mineral grains (Waite *et al.* 2004). Upon pressurization, the temperature immediately rose as much as 276.7 K owing to gas compression and hydrate formation. The initial formation subsided, and later was followed by rapid and then slow continued formation, with intermittent minor formation spikes. Hydrate was allowed to form over 28 h, and was still slowly forming when the next test began. On the basis of pressure and temperature measurements, approximately 63% of the water in the sample was converted to hydrate (assuming 100% cage occupancy).

3.1.4 Clathrates and Crystalline Clathrate Solids

Natural gas hydrates are solids that form from a combination of water and one or more hydrocarbon or nonhydrocarbon gases. In physical appearance, gas hydrates

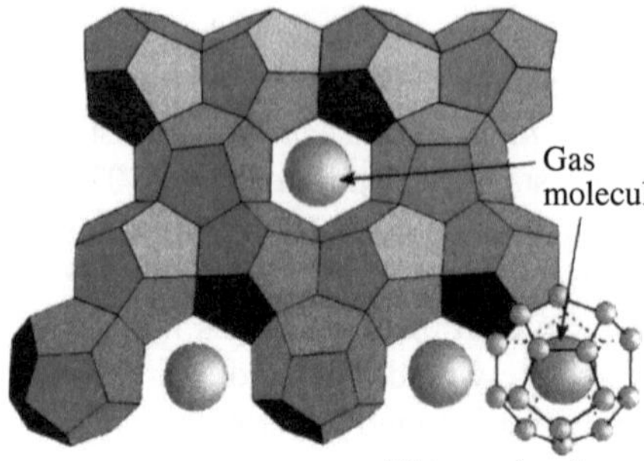

Fig. 3.2 Water-ice-like cage structure of gas hydrate

resemble packed snow or ice. In a gas hydrate, the gas molecules are "caged" within a crystal structure composed of water molecules (Fig. 3.2). Sometimes gas hydrates are called "gas clathrates." A gas hydrate, such as methane hydrate, is a crystalline solid known as a clathrate. The word "clathrate" has its origins in the Latin word *clatratus*, meaning "to enclose with bars." It follows then that clathrates are a class of chemical substances made of two unique materials, one of which encloses the other in an open, latticelike cage. There is no chemical bonding to hold the two materials together, only the physical structure. The most abundant naturally forming clathrate is methane hydrate.

Methane gas hydrate is a solid combination of methane and ice. It is found under continental shelves and on land under permafrost and can contain from 80 to 99.9% methane. Considerable knowledge has been accumulated on the occurrence of gas hydrates in nature and their properties. However, the issues of the technology and economics of natural gas production from gas hydrates are less well understood.

Clathrates occur wherever the conditions within the sediments are in the methane clathrate stability field, and where methane and water are available. This stability is limited by temperature and pressure: gas hydrates are stable at low temperatures and/or high pressures. The average thickness of the clathrate stability zone along continental margins is about 500 m. The depth of its lower boundary is determined by the geothermal gradient: at greater depths in the sediment, the temperature becomes too high for gas hydrates. The thickest clathrate zones occur in regions of low geothermal gradients; most clathrates occur within 2,000 m of the earth's surface (Hacisalihoglu *et al.* 2008).

The average hydrate composition is 1 mol of methane for every 5.75 mol of water. The observed density is around 0.9 kg/L. One liter of methane clathrate solid would contain 168 L of methane gas at standard temperature and pressure.

3.1.5 *Storage of Hydrogen in Clathrates*

Gas clathrate hydrates are crystalline inclusion compounds composed of a lattice of hydrogen-bonded water cages which can encage small guest molecules, such as methane, carbon dioxide, and hydrogen (Sloan and Koh 2008).

Storage of molecular hydrogen in clathrate hydrates has been achieved, with reversible release of hydrogen at ambient conditions. Hydrogen stored as a small guest molecule in different clathrate molecular compounds has been measured using Raman spectroscopy coupled with neutron diffraction.

Clathrate hydrates have been a scientific curiosity for many years owing to their intriguing nature. The detection in the mid-1930s of natural gas hydrates inside gas pipelines is a milestone marking the importance of hydrates to industry. Interest in gas hydrates comprises very broad areas, such as gas transmissions, water desalinization, and the exploitation of naturally occurring solid gas hydrate fields in the permafrost regions as alternative energy resources. Other areas of interest have been proposed for clathrate hydrates, namely, the deep-sea deposition of carbon dioxide clathrate and the possibility to store molecular hydrogen. The potential of these compounds to store hydrogen was first described in 2002 by Mao *et al.* (2002).

The storage of hydrogen in its molecular form in the clathrate cages offers some advantages that make hydrogen hydrates attractive. It is of great interest to stabilize the hydrogen hydrates at relatively low pressures and high temperatures to explore the possibility of using these solids as a safe hydrogen-storage method in the transportation sector. Furthermore, the storage of hydrogen in molecular form could lead to an increase in the overall energy efficiency as hydrogen is stored without the need for a chemical reaction for its storage and release.

The successful synthesis of hydrogen hydrates was a breakthrough in the development of materials for hydrogen storage. Very different from conventional hydrogen storage materials, the hydrogen hydrates are neither flammable nor corrosive, and provide a safe and environmentally friendly material to store hydrogen (Schuth 2005; Hu and Ruckenstein 2006).

Applications of hydrate technologies include the assessment of hydrates as a potential future energy source, energy storage, and industrial flow assurance. New insight into controlling hydrate formation and decomposition in these technological applications requires fundamental understanding of clathrate hydrate structural and physical properties, and crystal growth and decomposition processes.

3.2 Chemical Structures

3.2.1 Chemical Structure of the Water Molecule

Water has the chemical formula H_2O. Water as a liquid is the most common phase on the surface of the earth, followed by ice. The structure of water or that of molecules in water has largely been the province of chemists and geologists, and the reader must understand the differences in language and approach between these two communities.

According to molecular orbital theory, one molecule of water has two hydrogen atoms covalently bonded to a single oxygen atom by sp^3 hybrid orbitals (sp^3 hy-

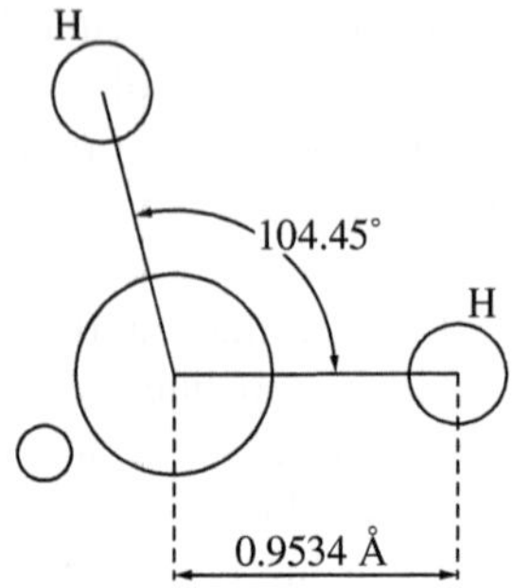

Fig. 3.3 The dimensions and geometric structure of a water molecule

brid orbitals are formed by the mixing of one *s* orbital and three *p* orbitals). The prominence of hydrogen bonding in the molecules is regularly commented on. Probably several hundred thousand papers have discussed the structure of the monomeric water molecule itself. Figure 3.3 shows the dimensions and geometric structure of a water molecule. The sp^3 hybrid orbitals are directed to the corners of a regular tetrahedron. The angle between any two orbitals is the tetrahedral angle 109.5°. Oxygen has only unpaired electrons, and hence it bonds with only two hydrogen atoms, which occupy two corners of a tetrahedron. The other two corners of the tetrahedron are occupied by unshared pairs of electrons. As experimentally measured, the H–O–H angle is 104.45°. Here there are two bulky unshared pairs of electrons compressing the bond angles. The O–H bond length is 0.9584 Å. It takes 118 kcal/mol to break one of the bonds of water.

Figure 3.4 shows the equilibrium vapor pressure curve for water. This is the partial pressure of water vapor as a function of temperature. The boiling point is the temperature where the vapor pressure reaches atmospheric pressure (760 mmHg). It is also called the normal boiling point.

Figure 3.5 shows a typical phase diagram for water. The dotted line gives the anomalous behavior of water. Water (liquid), ice (solid), and water vapor can

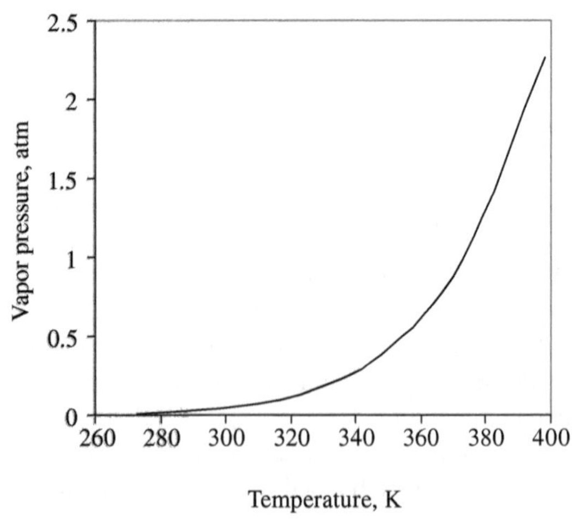

Fig. 3.4 Equilibrium vapor pressure curve for water

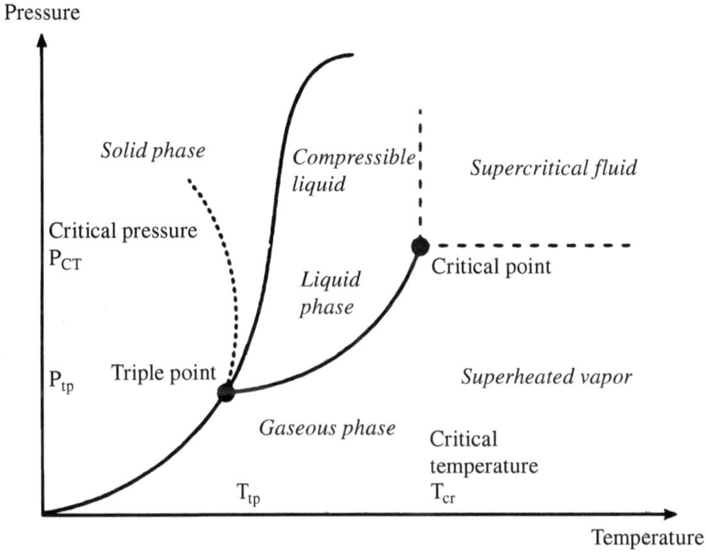

Fig. 3.5 A typical phase diagram for water

Table 3.3 Changes in different water properties at different temperatures

Property	Comment
Density	Maximum at 277 K
Refractive index	Thermal maximum near 273 K
Specific heat of water	Minimum at 308 K
Isothermal compressibility	Minimum at 323 K
Isothermal piezo-optic coefficient	Maximum near 323 K

coexist in a stable equilibrium at exactly 273.16 K and a partial vapor pressure of 0.006 atm. At that point, it is possible to change all of the substance to ice, water, or vapor by making arbitrarily small changes in pressure and temperature.

Table 3.3 shows the changes in different water properties at different temperatures. The density of liquid water is highest (1.00 kg/K) at 277 K. The specific heat of water is lowest at 308 K and the isothermal compressibility is lowest at 323 K.

3.2.2 Chemical Structure of the Methane Molecule

Figure 3.6 shows the tetrahedral structure and the space model structure of a methane molecule. Overlap of each of the sp^3 orbitals of carbon with an *s* orbital of hydrogen results in methane: carbon at the center of a regular tetrahedron, and the four hydrogens at the corners.

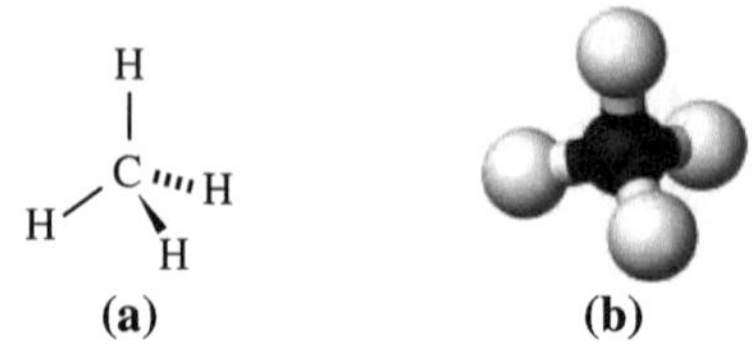

Fig. 3.6 Tetrahedral structure (a) and space model structure (b) of a methane molecule

Experimentally, methane has been found to have the highly symmetrical tetrahedral structure. Each C–H has exactly the same length, 1.10 Å; the angle between any pair of bonds is the tetrahedral angle 109.5°. It takes 104 kcal/mol to break one of the bonds of methane.

3.2.3 Crystalline Structure of Water Ice

As a naturally occurring crystalline solid, ice is considered to be a mineral consisting of hydrogen oxide. An unusual property of ice frozen at a pressure of 1 atm is that the solid is some 9% less dense than liquid water.

Figure 3.7 shows the crystalline structure of water ice. As water freezes, the molecules are forced to align themselves into a very particular structure – a hexagonal lattice that is the basis for the six-sided snowflakes. This shape also helps explain why water expands as it freezes, whereas all other solids contract. Ice takes up more space as a solid and is therefore less dense, meaning that ice floats in liquid water. No other compound behaves this way. Ice has a density of 0.9167 kg/L at 273 K, whereas water has a density of 0.9998 kg/L at the same temperature. Liquid water is densest, essentially 1.00 kg/K, at 277 K and becomes less dense as the water molecules begin to form the hexagonal crystals (as seen in Fig. 3.6) of ice as the temperature drops to 273 K. The density of ice increases slightly with decreasing temperature (the density of ice at 93 K is 0.9340 kg/L).

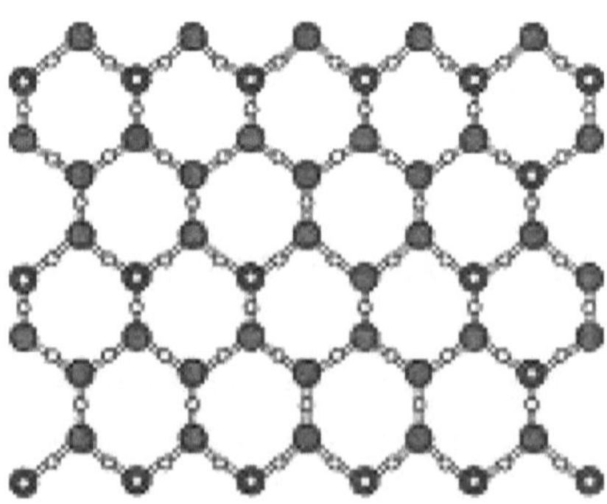

Fig. 3.7 Crystalline structure of water ice

3.2.4 Crystalline Structure of Gas Hydrate

The crystalline structure of hydrate is hydrogen-bonded by means of a nearly perfect tetrahedron. The range of motion of a methane molecule in a large cage is greater than that in a small cage. This implies that a methane molecule in a large cage more easily escapes from the hydrates than one in a small cage. Figure 3.8 shows the crystalline structure of methane gas hydrate.

Methane hydrate has a porous nature. Diffusion of gas and water through a hydrate has been observed at the grain level (Tohidi *et al.* 2001). When formed by rapidly cooling sediment, a hydrate preferentially grows at the gas–water interface, and a porous hydrate is formed. In an earlier study, the natural methane hydrate density was determined as 0.79 kg/L (Suess *et al.* 2002). The bulk densities of nearly 80 samples ranged from 0.35 to 0.75 kg/L and were inversely correlated with the pore volume, which ranged from 10 to 70% by volume. The density of methane hydrate is lower than the density of seawater, but the density of carbon dioxide hydrate is higher than the density of seawater.

Water crystallizes in the cubic system in clathrates, rather than in the hexagonal structure of normal ice. In this structure, the cages are arranged in body-centered packing; the unit cell contains 46 molecules of water and up to eight molecules of methane $[(CH_4)\cdot 5.75(H_2O)]$, but not all cages are occupied. Methane gas hydrate is formed when water molecules freeze around a molecule of methane gas. It is typically found in two distinct geological areas – on land in the subsurface sediments of permafrost regions, and in oceanic sediments under at least 450 m of water. Pressure, temperature, and the availability of water and methane are the determining factors in how much hydrate develops. When hydrates form, they fill in the pore space in the sediments. Hydrate formation draws methane molecules into closely packed lattice sites, effectively concentrating the methane. Methane hydrate is nonstoichiometric in that the crystal structure of the hydrate can be established without all the lattice sites being occupied.

Table 3.4 shows the physical properties of water ice and methane hydrate (Sloan 1998). As can be seen from Table 3.4, there are important differences between the physical properties of water ice and methane hydrate. Figure 3.9 shows a typical pressure–temperature diagram for the water–methane gas hydrate system.

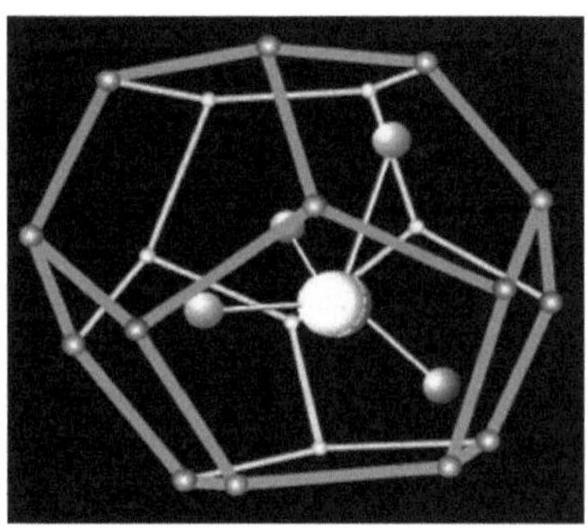

Fig. 3.8 An example of the crystalline structure of methane gas hydrate

Table 3.4 Physical properties of water ice and methane hydrate

Property	Ice	Hydrate
Dielectric constant at 273 K	94	~58
NMR rigid lattice 2nd moment of H_2O protons (G2)	32	33 ± 2
Water molecule reorientation time at 273 K (μs)	21	~10
Diffusional jump time of water molecules at 273 K (μs)	2.7	>200
Isothermal Young's modulus at 268 K (109 Pa)	9.5	~8.4
Pressure wave velocity (km/s)	3.8	~4
Transit time (μs/ft)	3.3	92
Velocity ratio Vp/Vs (272 K)	1.88	1.95
Poisson ratio	0.33	~0.33
Bulk modulus (272 K)	8.8	5.6
Shear modulus (272 K)	3.9	2.4
Bulk density (kg/L)	0.916	0.912
Adiabatic bulk compressibility at 273 K and 10–11 Pa	12	~14
Thermal conductivity at 273 K (W/m K)	2.25	0.50 ± 0.02
Heat of fusion (kJ/mol)	6	54

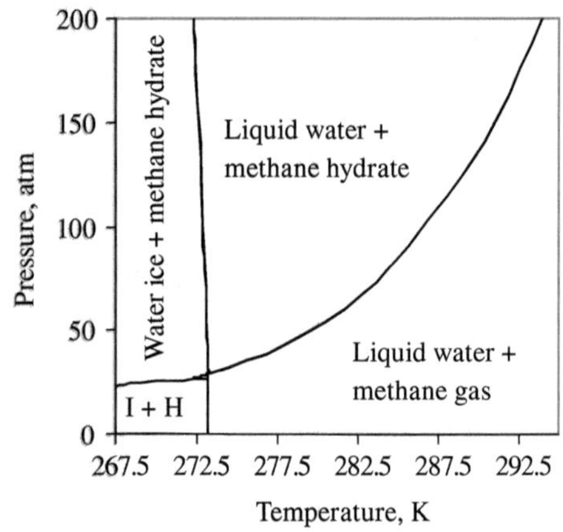

Fig. 3.9 A typical pressure–temperature diagram for the water–methane gas hydrate system. *I+H* water ice plus methane hydrate

3.3 Crystal Types of Gas Hydrates

Clathrate ices form from water and nonstoichiometric amounts of small nonpolar molecules under moderate pressure (typically a few megapascals) and at cold temperatures (typically close to 273 K, but increased pressure raises the melting point). Their structures require a minimum amount of small molecules to fit into and stabilize the cavities without forming any covalent or hydrogen bonds with the water molecules. Without these interstitial molecules, the clathrate cavities would collapse at positive pressures and they have been shown to dissipate, if surprisingly slowly, after the clathrate ice melts (Buchanan *et al.* 2005). During formation and dissociation, the solid clathrates interact significantly with the structure of the neighboring aqueous solution (Gao *et al.* 2005).

Table 3.5 Types of clathrates

Type	Lattice	Space group	Unit cell	Unit cell formula[a]
Clathrate sI	Cubic	*Pm3n*	$a = 1.20$ nm	$(S)_2(L)_6 \cdot 46H_2O$
Clathrate sII	Face-centered cubic	*Fd3m*	$a = 1.73$ nm	$(S)_{16}(L+)_8 \cdot 136H_2O$
Clathrate sH	Hexagonal	*P6/mmm*	$a = 1.23$ nm $c = 1.02$ nm	$(S)_5(L{+}{+}) \cdot 34H_2O$

sII structure II, *sH* structure H, *S* small guest, *L* large guest, *L+* larger guest, *L++* largest guest
[a] Not all cavities would normally be filled.

Some clathrate hydrates can form, at atmospheric pressure, at the interface between a liquid of suitable guest molecules and water, e.g., CH_3CCl_2F in clathrate structure II (sII) hydrate (Ohmura *et al.* 1999). At atmospheric pressure, most clathrate hydrates decompose to release the guest molecules, except at low temperatures (e.g., below 270 K), where they may remain in a metastable state for several hours. At very high pressures, clathrate hydrates show complex phase behavior, often giving filled hexagonal ice (Hiraia *et al.* 2004) with the smaller guest molecules/atoms, then at higher pressures they break down to give high-density ice and a solid phase formed by the guest molecules (Sanloup *et al.* 2002). Gas hydrates have been recently reviewed (Koh 2002). Water itself cannot be contained in the cavities of solid clathrates (McCarthy and Jordan 2006).

Depending on the size of the guest molecule, natural gas hydrates can consist of any combination of three crystal structures (Table 3.5): (1) structure I (sI), (2) sII, and (3) structure H (sH). When pure liquid water freezes, it crystallizes with hexagonal symmetry, but when it "freezes" as a hydrocarbon hydrate, it does so with cubic symmetry for sI and sII, reverting to hexagonal symmetry for sH. Table 3.6 shows the characteristic properties of the clathrates.

Table 3.6 Characteristic properties of the clathrates

	Cavity				
	5^{12}	$5^{12}6^2$	$5^{12}6^4$	$5^{12}6^8$	$4^35^66^3$
No. of H_2O molecules	20	24	28	36	20
Mean cavity radius (Å)	3.95	4.33	4.73	5.71	4.06
Free volume (Å^3)	51	77	120	213	44
No. of clathrate sI/ unit cell	2	6	–	–	–
No. of clathrate sII/ unit cell	16	–	8	–	–
No. of clathrate sH/ unit cell	3	–	–	1	2
Guest molecules Approximate radius of guest molecules (Å)	Ar, O_2, N_2, CH_4 1.8–2.2	CO_2, C_2H_6 1.8–2.7	C_3H_8, $(CH_3)_3CH$ 2.8–3.1	$(CH_3)_3CC_2H_5$ 3.5–4.3	CH_4 1.8

3.3.1 Clathrate Structure I

The sI gas hydrates contain 46 water molecules per unit cell arranged in two dodecahedral voids and six tetrakaidecahedral voids (the water molecules occupy the apices in the stick diagrams of the void types shown in Fig. 3.10), which can accommodate at most eight guest molecules up to 5.8 Å in diameter. The inclusion of both methane and ethane but not propane is allowed by sI. The sI gas hydrates are usually formed by smaller molecules such as methane, ethane, and carbon dioxide, 46 water molecules per eight gas molecules, and consist of eight pentagonal dodecahedron cages.

Figure 3.10 shows the cubic clathrate sI network formed by small nonpolar (gaseous) molecules, such as methane and carbon dioxide, in aqueous solution [e.g., $(CO_2)_{8-y}{\cdot}46H_2O$] under pressure and at low but not necessarily (normally) freezing temperatures (only the oxygen atoms of water are shown). The included molecules randomly occupy many of the cavities depending on their size. Linear tetrakaidecahedral ($5^{12}6^2$) cavities form three orthogonal axes holding a dodecahedral cavity wherever they cross (ratio 6:2 per unit cell); each dodecahedral cavity sits (in a body-centered-cubic arrangement) within a cube formed by six tetrakaidecahedral ($5^{12}6^2$) cavities. These $5^{12}6^2$ cavities join at their hexagonal faces to form columns.

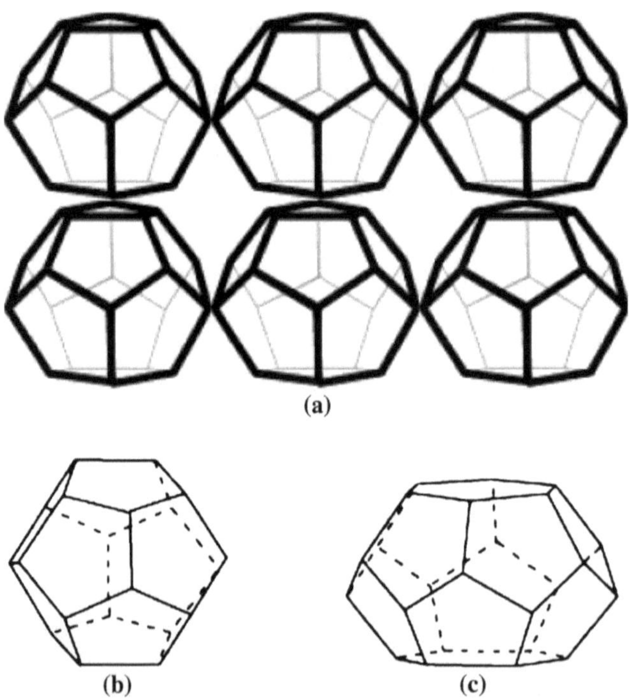

Fig. 3.10 Structure I gas hydrate crystal structures: (a) general, (b) pentagonal dodecahedron, and (c) tetrakaidecahedron

About 6.4 trillion tons of methane lies at the bottom of the oceans in the form of its clathrate hydrate (Buffett and Archer 2004). Each kilogram of fully occupied hydrate (actually only about 96% occupancy is found) holds about 187 L of methane (at atmospheric pressure).

3.3.2 Clathrate Structure II

The sII gas hydrates contain 136 water molecules per unit cell arranged in 16 dodecahedral voids and eight hexakaidecahedral voids, which can also accommodate up to 24 guest molecules, but to a larger diameter of 6.9 Å. This allows inclusion of propane and isobutane in addition to methane and ethane. The sII gas hydrates are usually formed by larger molecules such as propane and isobutane, 136 water molecules per 24 gas molecules, and consist of 24 hexakaidecahedron cages.

Figure 3.11 shows the sII hydrate structure (cubic crystals containing sixteen 5^{12} cavities, eight larger $5^{12}6^4$ cavities, and 136 H_2O molecules per unit cell, and

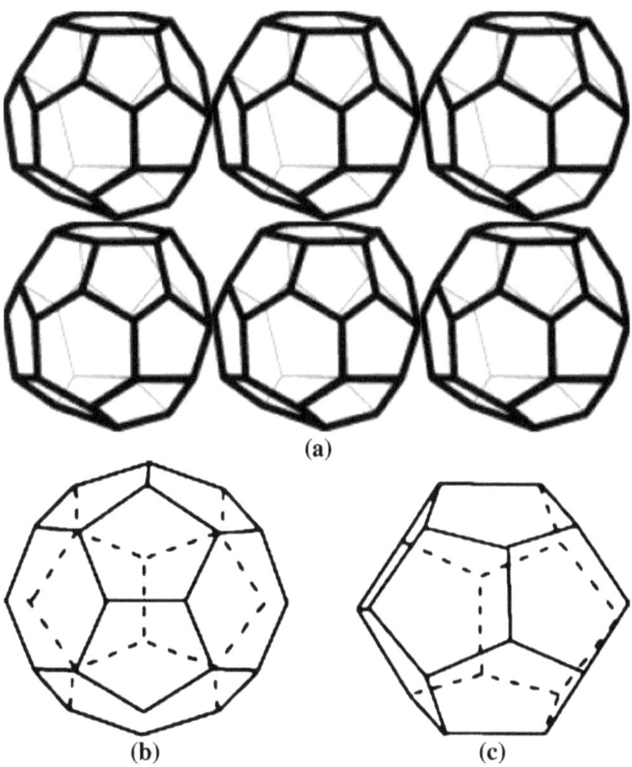

Fig. 3.11 Structure II gas hydrate crystal structures: (a) general, (b) hexakaidecahedron, and (c) dodecahedron

containing larger molecules such as 2-methylpropane in the larger cavities only). The tetrahedral $5^{12}6^4$ cavities form an open tetrahedral network, with their centers arranged in a manner reminiscent of the cubic ice structure and separated by groups of three 5^{12} cavities. The large proportion of 5^{12} cavities is thought to be responsible for the similarities in the Raman spectra to gas-saturated water (Schicks *et al.* 2005).

Rather surprisingly, the sII clathrate forms with molecular hydrogen (H_2), four molecules sitting in the large cages and one molecule (Katsumasa *et al.* 2007) or two molecules (Patchkovskii and Tse 2003) in the small cages, that is, $(2H_2)_{16}(4H_2)_8 \cdot 136H_2O$ (Katsumasa *et al.* 2007).

A unit cell (the smallest repeatable element) of an sI hydrate consists of 46 water molecules surrounding two small cavities and six medium-sized cavities. The unit cell of sII hydrates consists of 136 water molecules creating 16 small cavities and eight large cavities. Both sI and sII can be stabilized by filling at least 70% of the cavities by a single guest gas – and are therefore known as simple hydrates.

3.3.3 Clathrate Structure H

The sH gas hydrates, which contain 34 water molecules per unit cell arranged in three pentagonal dodecahedral voids, two irregular dodecahedral voids, and one icosahedral void, can accommodate even larger guest molecules, such as isopentane. The sH gas hydrates are usually formed by large molecules such as methylcyclohexane, but only in the presence of a smaller molecule, 34 water molecules per six gas molecules. The large molecule occupies the larger cage, and the small molecule occupies the smaller cage.

Figure 3.12 shows the sH hydrate structure. It has hexagonal crystals containing three 5^{12} cavities, two small $4^35^66^3$ cavities, one large $5^{12}6^8$ cavity, and 34 water molecules per unit cell, and contains even larger molecules such as 2,2-dimethylbutane in the larger cavities only. Each $5^{12}6^8$ barrel-shaped cavity is surrounded by six $4^35^66^3$ cavities around its central ring of six hexagons. These $5^{12}6^8$ cavities join at their top and bottom hexagonal faces to form columns.

The hydrocarbon hydrates are nonstoichiometric substances, i.e., their compositional proportions are not fixed. A variable number of guest molecules up to the maximums given above can be accommodated in the host lattice since not all of the available lattice positions need be filled. Typically, the volume of gas included in a fixed volume of hydrate increases in response to either lower temperature or higher pressure. Thus, given the substantial density difference between water and free gas, 1 vol of water can accommodate from 70 to over 160 vol of gas depending on how many of the available voids are filled (the degree of saturation). Natural gas hydrates are often undersaturated, with most samples of the simplest and most common sI type falling in the 70–90% saturated range.

There are three experimental methods for hydrate formation in saturated conditions:

1. A saturated specimen has gas injected into it until a predetermined quantity of water has been pushed out as a measure of hydrate content.
2. A dry specimen is filled with gas to a certain pressure, with water then injected to drive the pore pressure up into the hydrate stability zone and saturate the specimen.
3. Hydrate is formed from the dissolved gas phase.

Although making hydrate by the first method has been successfully implemented by Winters *et al.* (2004), Stoll and Bryan (1979), and Brewer *et al.* (1997), the solubility of methane in water is low and pressure-dependent. Therefore, hydrate formation from dissolved methane gas would create a maximum of 4% hydrate in the pore space, given a maximum cell pressure of 25 MPa. Tohidi *et al.* (2001) and Buffett and Zatsepina (2000) have shown that carbon dioxide can be used to successfully form hydrate from solution.

Methane hydrate has a porous nature. Diffusion of gas and water through hydrate has been observed at the grain level (Tohidi *et al.* 2001). When formed by rapidly cooling a sediment, hydrate preferentially grows at the gas–water interface,

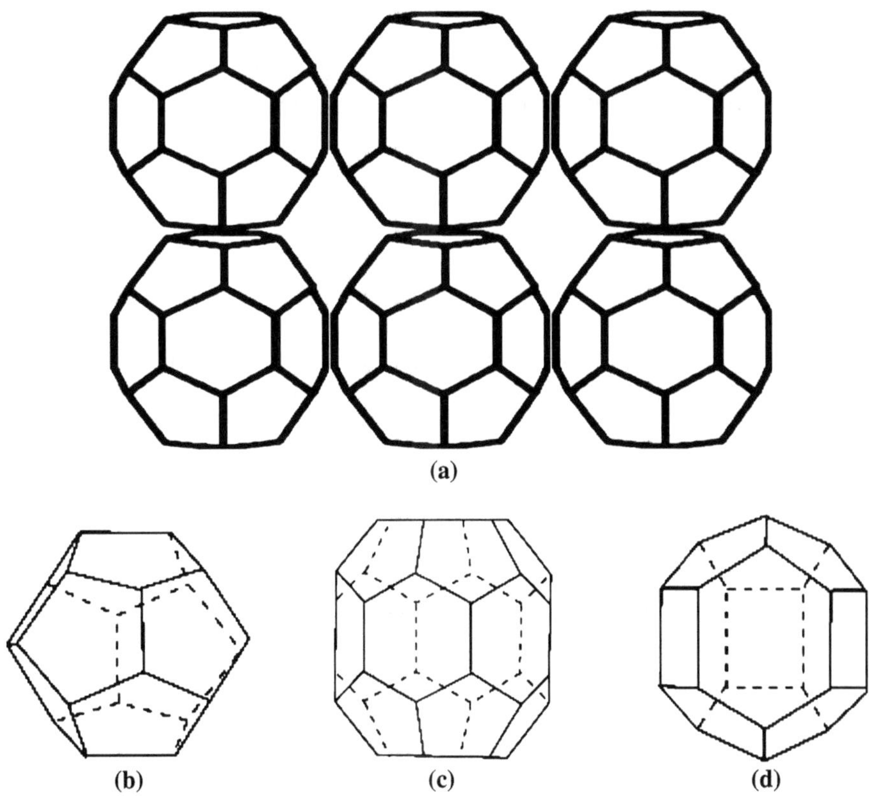

Fig. 3.12 Structure H gas hydrate crystal structures: (a) general, (b) pentagonal dodecahedron, (c) icosahedron, and (d) irregular dodecahedron

and porous hydrate is formed. During the passing of a seismic wave, this porous hydrate is likely to deform and squirt flow through the hydrate grains themselves may occur. Squirt flow may also develop between hydrate grains and sand grains, where hydrate rests at grain boundaries. The increase in hydrate content will therefore provide more conduits for water movement, and so attenuation increases.

The possible occurrence of syntectonic dissociation of methane hydrate to methane plus free water in these experiments suggests that the high strength measured may be only a lower bound. On Earth, high strength in hydrate-bearing formations implies higher energy release upon decomposition and subsequent failure. In the outer solar system, if Titan has a 100-km-thick near-surface layer of high-strength, low-thermal-conductivity methane hydrate as has been suggested, its interior is likely to be considerably warmer than previously expected. Dispersed gas hydrates in marine sediments at typical volume concentrations of less than 5%, on the other hand, will probably not have a direct rheological affect. Other properties of gas hydrates, such as cohesiveness with sediment particles, and more importantly the overall phase stability, probably have a greater influence on key formation properties such as slope stability. At the other extreme, the presence of a moonwide, 100-km-thick layer of very strong, thermally insulating methane hydrate must have a first-order effect on the evolution of Titan. Controlled experiments on synthetic hydrate–sediment aggregates are clearly called for to determine what roles gas hydrate rheological properties may play in governing the inelastic behavior of natural sediment–hydrate aggregates (Zhang *et al.* 1999; Durham *et al.* 2003).

3.4 Methane Gas Fuel

Methane was discovered and isolated by Alessandro Volta between 1776 and 1778 when he was studying marsh gas from Lake Maggiore. It is the dangerous firedamp of the coal mine, and can be seen as marsh gas bubbling to the surface of swamps.

Methane is an end product of the anaerobic decay of plants, that is, of the breakdown of certain very complicated molecules. As such, it is the major constituent (up to 97%) of natural gas. If methane is wanted in very pure form, it can be separated from the other constituents of natural gas by fractional distillation.

Methane is a chemical compound with the molecular formula CH_4. It is the simplest saturated hydrocarbon (alkane), and the principal component of natural gas, biogas, landfill gas, and methane hydrate. Methane has a regular tetrahedron molecular structure. Methane's bond angles are 109.5°.

Methane is colorless and, when liquefied, is less dense than water (specific gravity 0.4). Because the methane molecule is highly symmetrical, the polarities of the individual C–H bonds cancel out; as a result, the molecule itself is nonpolar. In agreement with the rule of thumb that "like dissolves like," it is only slightly soluble in water, but is very soluble in organic liquids such as gasoline, diethyl ether, and ethanol.

Table 3.7 Chemical compositions of some fuels and pure compounds

Sample	C	H	O	N	S
No. 1 diesel fuel	85.9	13.9	0.11	0.15	0.04
No. 2 diesel fuel	86.0	13.8	0.06	0.10	0.04
Methane	74.9	25.1	–	–	–
n-Butane	82.9	17.1	–	–	–
n-Pentane	83.2	16.8	–	–	–
Ethylene	85.6	13.4	–	–	–
Benzene	92.3	7.7	–	–	–
Acetylene	92.3	7.7	–	–	–
Naphthalene	93.7	6.3	–	–	–
Ethyl alcohol	52.1	13.2	34.7	–	–
Propane	81.7	18.3	–	–	–
Cyclopropane	85.6	14.4	–	–	–
Cyclobutane	85.6	14.4	–	–	–
Cyclopentane	85.6	14.4	–	–	–
Wood bark	53.0	6.1	40.6	0.2	<0.01
Solid waste	48.3	5.7	45.3	0.7	<0.01
Corncob	49.0	5.4	44.6	0.4	<0.01

Attraction between such nonpolar molecules is limited to van der Waals forces; for such small molecules, these attractive forces must be tiny. It is not surprising, then, that these attractive forces are easily overcome by thermal energy, so melting and boiling occur at very low temperatures: melting point 90 K, boiling point 111.5 K. As a consequence, methane is a gas at ordinary temperatures.

Burning methane in the presence of oxygen produces carbon dioxide and water:

$$CH_4 + 2O_2 \rightarrow CO_2 + 2H_2O + 891\,\text{kJ/mol}. \tag{3.4}$$

The relative abundance of methane and its clean-burning process make it a very attractive fuel. However, because it is a gas at normal temperature and pressure, methane is difficult to transport from its source. In its natural-gas form, it is generally transported in bulk by pipeline or LNG carriers; few countries still transport it by truck.

Table 3.7 shows the chemical compositions of some fuels and pure compounds. Examination of the data for a great many compounds has shown that the heat of combustion of an aliphatic hydrocarbon agrees rather closely with that calculated by assuming a certain characteristic contribution from each structural unit. For straight-chain alkanes, the contribution of each methylene group, $-CH_2-$, to the heat of combustion is about 658.7 kJ/mol. For cyclopropane, the heat of combustion per $-CH_2-$ group is 37.7 kJ/mol higher than the open-chain value of 658.7 kJ/mol; for cyclobutane, it is 29.3 kJ/mol higher than the straight-chain value. Cyclopropane and cyclobutane produce more energy per $-CH_2-$ group than a straight-chain compound. Then, cyclopropane and cyclobutane are less stable

Table 3.8 Average bond energies for bonds in polyatomic molecules

Bond	Bond energy		
	kcal/mol	kcal/g	MJ/kg
C–C	83	3.455	14.460
C=C	146	6.078	25.435
C≡C	199	5.523	23.113
C–O	85	3.035	12.699
C=O	170	6.069	25.399
O–H	111	6.526	27.313
O=O	119	3.719	14.517
N≡N	227	8.107	33.928
C≡O	257	9.175	38.397
$CH_4 \rightarrow CH_3 + H$	104	6.487	27.148
$CH_3 \rightarrow CH_2 + H$	106	7.055	29.527
$CH_2 \rightarrow CH + H$	106	7.563	31.651
$CH \rightarrow C + H$	81	6.227	26.060
C–H (average)	98	7.534	31.529
H–H	104	51.587	215.893

than straight-chain compounds (Morrison and Boyd 1983). Among the fuels including carbon and hydrogen, methane has the highest hydrogen content.

Average bond energies for bonds in polyatomic molecules are given in Table 3.8. For a molecule to break into atoms, an equivalent amount of energy must be consumed. The amount of energy consumed or liberated when a bond is broken or formed is known as the bond dissociation energy. The bond dissociation energy is characteristic of the particular bond (Morrison and Boyd 1983). It is possible to measure the bond dissociation energies in polyatomic molecules as well. Systematic study of many compounds containing the same bond enables us to derive an average value for this bond. The variation in a given bond energy from compound to compound is about ±10%.

3.4.1 Combustion of Methane

Combustion is a basic chemical process that releases energy from a fuel and air mixture. For combustion to occur, fuel, oxygen, and heat must be present together. The chemical reaction of inflammable matter with oxygen from the air is accompanied by heat release. The quantity of heat evolved when 1 mol of a hydrocarbon is burned to produce carbon dioxide and water is called the heat of combustion. Combustion producing carbon dioxide and water is characteristic of organic compounds; under special conditions it is used to determine their carbon and hydrogen content. During combustion, the combustible part of the fuel is subdivided into a volatile part and a solid residue. During heating, evaporates together with some carbon in the form of hydrocarbon combustible gases and carbon monoxide are released by thermal degradation of the fuel. Carbon monoxide is mainly formed (1) from reduction of carbon dioxide with unreacted carbon

$$CO_2 + C \rightarrow 2CO \tag{3.5}$$

and (2) from degradation of carbonyl fragments (–CO) in the fuel molecules at 600–750 K.

The combustion process is started by heating the fuel above its ignition temperature in the presence of oxygen or air. Under the influence of heat, the chemical bonds of the fuel are cleaved. If complete combustion occurs, the combustible elements (C, H, and S) react with the oxygen in the air to form CO_2, H_2O, and mainly SO_2.

If not enough oxygen is present or the fuel and air mixture is inadequate, then the burning gases are partially cooled below the ignition temperature and the combustion process is incomplete. The flue gases then still contain combustible components, mainly carbon monoxide, unburned carbon, and various hydrocarbons (C_xH_y).

The standard measure of the energy content of a fuel is its heating value, sometimes called the calorific value or heat of combustion. In fact, there are multiple values for the heating value, depending on whether it measures the enthalpy of combustion (ΔH) or the internal energy of combustion (ΔU), and whether for a fuel containing hydrogen, product water is accounted for in the vapor phase or the condensed (liquid) phase. With water in the vapor phase, the lower heating value at constant pressure measures the enthalpy change due to combustion (Jenkins et al. 1998). The heating value is obtained from the complete combustion of a unit quantity of solid fuel in an oxygen-bomb calorimeter under carefully defined conditions. The gross heat of combustion, or higher heating value, is obtained by the oxygen-bomb-calorimeter method as the latent heat of moisture in the combustion products recovered.

Combustion of methane is the principal reaction that takes place during the burning of natural gas. It is hardly necessary to emphasize its importance in the areas where natural gas is available; the important product is not CO_2 or H_2O, but heat.

In the combustion of methane, several steps are involved. Methane is believed to form formaldehyde (H_2CO). The formaldehyde gives a formyl radical ($HCO\cdot$), which then forms carbon monoxide. The process is called oxidative pyrolysis:

$$CH_4 + O_2 \rightarrow CO + H_2 + H_2O \tag{3.6}$$

Following oxidative pyrolysis, the H_2 is oxidized to H_2O. This occurs very quickly, usually in significantly less than 1 ms:

$$2H_2 + O_2 \rightarrow 2H_2O. \tag{3.7}$$

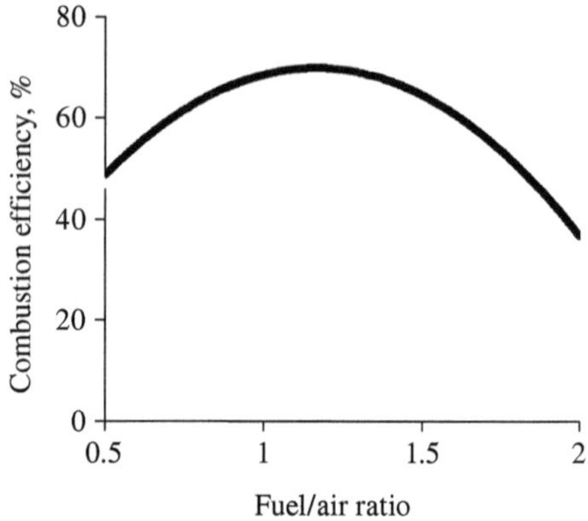

Fig. 3.13 A typical diagram for combustion efficiency

Finally, the CO is oxidized to CO_2 and more heat is released. This process is generally slower than the other chemical steps, and typically requires a few milliseconds to occur:

$$2CO + O_2 \rightarrow 2CO_2. \tag{3.8}$$

The following total equation results from the aforementioned reactions:

$$CH_4(g) + 2O_2(g) \rightarrow CO_2(g) + 2H_2O(l) + 891\,kJ/mol. \tag{3.9}$$

Compared with burning other hydrocarbon fuels, burning methane produces less carbon dioxide for each unit of heat released. Methane's combustion heat is lower than that of any other hydrocarbon; but the ratio of the molecular mass (16.0 g/mol) divided by the heat of combustion (891 kJ/mol) shows that methane, being the simplest hydrocarbon, produces more heat per unit mass than other complex hydrocarbons.

Methane is the major component of natural gas, about 87% by volume. At ambient temperature and standard pressure, methane is a colorless, odorless gas; the smell characteristic of natural gas as used in homes is an artificial safety measure caused by the addition of an odorant, often methanethiol or ethanethiol. The normal boiling point of methane is 112 K at a pressure of 1 atm. As a gas, it is flammable only over a narrow range of concentrations (5–15%) in air. Liquid methane does not burn unless subjected to high pressure (normally 4–5 atm).

In general, combustion efficiency as defined in flue gas analysis standards is simply reduced by the stack loss. Combustion efficiency is based on the flue gas temperature and inlet air temperature. Combustion efficiency calculations assume complete fuel combustion and are based on the following three factors: (1) the chemistry of the fuel, (2) the net temperature of the stack gases, and (3) the percentage of oxygen or carbon dioxide by volume after combustion.

Combustion efficiency relates to the part of the reactants that combine chemically. It increases with increasing temperature of the reactants, increasing time that the reactants are in contact, increasing vapor pressures, increasing surface areas, and increasing stored chemical energy.

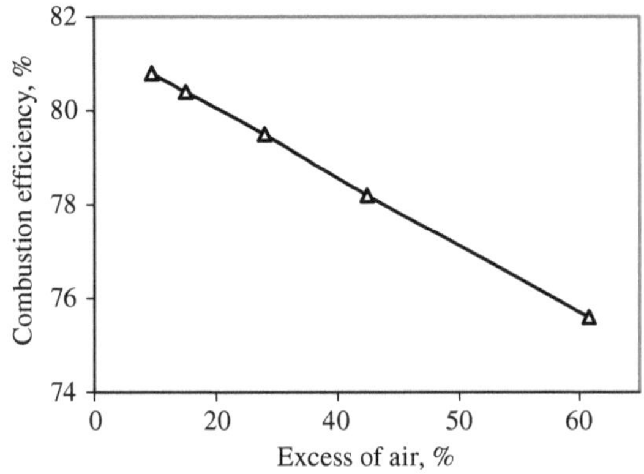

Fig. 3.14 Typical combustion efficiency versus excess of air for methane gas

Figure 3.13 shows a typical diagram for combustion efficiency. Without enough combustion air, the combustion efficiency is low. The combustion efficiency increases with the fuel-to-air ratio. The combustion efficiency is lower than the maximum value in a stoichiometric fuel/air mixture. Figure 3.14 shows the typical combustion efficiency versus excess of air for methane gas.

3.4.2 Main Reactions of Methane

The main reactions with methane are combustion, steam reforming to form syngas, and halogenation. In general, methane reactions are hard to control. Hydrogen gas can be obtained by direct thermolysis of methane and pyrolysis of hydrogen sulfide. Direct thermal dissociation of methane and hydrogen sulfide does not generate greenhouse gases. However, compared with the steam methane reforming (SMR) process, thermolysis of methane and hydrogen sulfide generates lesser amounts of hydrogen per mole of methane and hydrogen sulfide reacted. The reaction of hydrogen sulfide with methane can be thought of as the sulfur analog of the SMR process, as indicated by the following two reactions (Hacisalihoglu *et al.* 2008):

$$H_2S \rightarrow H_2 + 0.5S_2 \qquad \Delta H = +79.9\,\text{kJ/mol}, \tag{3.10}$$

$$CH_4 + 2S_2 \rightarrow CS_2 + 2H_2S \qquad \Delta H = -107\,\text{kJ/mol}. \tag{3.11}$$

The overall reaction may be written as follows:

$$CH_4 + 2H_2S \rightarrow CS_2 + 4H_2 \qquad \Delta H = +232.4\,\text{kJ/mol}. \tag{3.12}$$

The by-product obtained from the overall reaction is carbon disulfide (CS_2).

The reaction between methane and hydrogen sulfide given by Eq. 3.8 is the well-known methane process for production of carbon disulfide. Most commercial methane–sulfur processes employ a silica gel/aluminum catalyst for carbon disulfide production. The reaction of methane with sulfur is thermodynamically favorable for carbon disulfide formation, and conversion is usually in the range of 90–95% with respect to methane (UEIC 1989). The industrial methane–sulfur process operates in the temperature range of 775–925 K and the pressure range of 4–7 atm.

Through controlled partial oxidation of methane and the high-temperature catalytic reaction with water, methane is an increasingly important source of products other than heat: of hydrogen, used in the manufacture of ammonia; of a mixture of carbon monoxide and hydrogen, used in the manufacture of methanol and other alcohols; and of acetylene, itself the starting point for large-scale production of many organic compounds.

Under the influence of ultraviolet light or at a temperature of 525–675 K, a mixture of methane and chlorine reacts vigorously to yield hydrogen chloride (HCl) and methyl chloride (CH_3Cl). This reaction is called chlorination. Chlorination is a typical example of a broad class of organic reactions known as substitution.

$$CH_4 + Cl_2 + \text{light or heat} \rightarrow CH_3Cl + HCl \tag{3.13}$$

The methyl chloride can itself undergo further substitution to form more hydrogen chloride and methylene chloride (CH_2Cl_2):

$$CH_3Cl + Cl_2 + \text{light or heat} \rightarrow CH_2Cl_2 + HCl. \tag{3.14}$$

In a similar way, chlorination may continue to yield chloroform ($CHCl_3$) and carbon tetrachloride (CCl_4). Because of the great difference in their boiling points, it is easy to separate the excess methane from the methyl chloride and the other products.

The strength of the C–H covalent bond in methane is among the strongest in all hydrocarbons, and thus the use of methane as a chemical feedstock is limited. Despite the high activation barrier for breaking the C–H bond, methane is still the principal starting material for manufacture of hydrogen in steam reforming. SMR to form syngas is represented by the following equation:

$$CH_4 + H_2O \leftrightarrows CO + 3H_2. \tag{3.15}$$

Methane is a favorable source to produce hydrogen. For example, hydrogen can be produced from SMR, catalytic partial oxidation of methane, carbon dioxide reforming of methane, oxy-steam reforming of methane, and combined water and carbon dioxide reforming of methane over nickel and precious metal catalysts (Peña *et al.* 1996; Roh *et al.* 2002; Bradford and Vannice 1999; Koo *et al.* 2008a, b). Among the various types of reforming reactions, SMR is a well-established process to produce hydrogen for fuel cells (Rostrup-Nielsen 1984; Peña *et al.* 1996). However, SMR is a highly energy intensive process owing to its high endothermicity. Thus, high temperature is favorable for SMR because methane is a thermodynamically very stable molecule with a high C–H bond dissociation energy (104 kcal/mol) (Choudhary and Goodman 2000). A low-temperature SMR reaction can be carried out at atmospheric pressure in a fixed-bed reactor system.

In an earlier study (Laosiripojana and Assabumrungrat 2007), methane with steam was directly fed to a nickel–yttrium-stabilized zirconia cermet (Ni/YSZ) anode of a direct internal reforming solid oxide fuel cell (DIR-SOFC). It was found that methane with appropriate steam content can be directly fed to a Ni/YSZ anode without the problem of carbon formation, and methanol can also be introduced at a temperature as high as 1,275 K. A solid oxide fuel cell (SOFC) is an electrochemical energy conversion unit that converts chemical energy to electric energy and heat with greater energy efficiency and lower pollutant emission than the combustion process (Minh and Takahashi 1995).

It is well established that hydrogen and carbon monoxide can be typically used as the fuels for a SOFC. Furthermore, as a SOFC is generally operated at a high temperature (975–1,375 K), methane can also be directly used as fuel instead of hydrogen and carbon monoxide by feeding it straight to the anode side of the SOFC; this operation is called a DIR-SOFC. The advantage of the DIR-SOFC is that the hydrogen consumption by the electrochemical reaction can directly promote the reforming or conversion of hydrocarbons at the anode side. Therefore, the DIR-SOFC results in high conversion and high efficiency (Laosiripojana and Assabumrungrat 2007).

Ni/YSZ is the most common SOFC anode material owing to its cost-effectiveness compared with other supported metals (e.g., Co, Pt, Ru, and Rh) and also because it fits well with fuel cell design requirements. In addition, this material also provides catalytic reforming activity, which is beneficial for the DIR-SOFC operation. The nickel content of the Ni/YSZ anode is usually 40–60% in order to match the thermal expansion of yttrium-stabilized zirconia. Some previous researchers have investigated the performance of DIR-SOFC operation fueled by methane. Yentekakis *et al.* (1993) investigated the effect of steam on the SMR rate over Ni/YSZ in the temperature range from 1,073 to 1,203 K by varying the ratio of H_2O to CH_4 from 0.15 to 2.0. The experiment indicated the strong influence of steam on the reforming rate, which could be due to the steam deficiency. In contrast, Achenbach and Riensche (1994) reported no influence of inlet steam partial pressure on the rate of SMR over nickel cermet (20 wt% Ni and 80 wt% ZrO_2) at 973–1,213 K with the inlet H_2O to CH_4 ratio from 2.6 to 8.0. The tests showed that, within the accuracy of the data, no effect of the H_2O partial pressure on the catalytic reforming process could be observed. Owing to the high conversion rates of CH_4 at high temperatures, however, mass-transfer effects occurred, which must be taken into account when evaluating the steam-reforming data. Dicks *et al.* (2000) observed that the dependence of the rate of SMR over a Ni/ZrO_2 anode was a function of both temperature and gas compositions. For the purposes of optimizing important system parameters in DIR-SOFC systems, detailed knowledge of the rate of SMR on the anode is needed.

Table 3.9 shows the yields of products from SMR over Ni/Ce–ZrO_2 plus Ni/YSZ at different temperatures (inlet CH_4 to H_2O ratio of 1:3). Table 3.10 shows the yields of products from the steam reforming of methane over Ni/Ce–ZrO_2 plus Ni/YSZ at different CH_4 to H_2O ratios at 1,175 K.

Before modern production technologies were developed in the 1920s, methanol was obtained from wood as a coproduct of charcoal production and, for this reason, was commonly known as wood alcohol. Methanol is currently manufactured

Table 3.9 Yields of products from the steam reforming of methane over Ni/Ce–ZrO_2 plus nickel–yttrium-stabilized zirconia cermet (Ni/YSZ) at different temperatures (inlet CH_4 to H_2O ratio of 1:3)

Temperature (K)	Hydrogen (%)	Carbon monoxide (%)	Carbon dioxide (%)
975	55.3	26.1	20.6
1,025	60.0	24.3	15.7
1,075	67.6	20.9	11.5
1,125	72.9	20.5	6.6
1,175	76.1	17.4	6.5
1,200	78.7	15.4	5.9
1,225	81.3	14.2	4.5
1,250	84.5	11.8	3.7
1,275	87,6	9.9	2.5

worldwide by conversion or is derived from syngas, natural gas, refinery off-gas, coal, or petroleum:

$$2H_2 + CO \rightarrow CH_3OH. \tag{3.16}$$

The chemical compositions of syngas obtained from coal and from natural gas can be identical, with the same H_2 to CO ratio. A variety of catalysts are capable of causing the conversion, including reduced NiO-based preparations, reduced Cu/ZnO shift preparations, Cu/SiO_2 and Pd/SiO_2, and Pd/ZnO (Takezawa *et al.* 1987; Iwasa *et al.* 1993).

Methanol is currently made from natural gas, but can also be made using biomass via partial oxidation reactions. Biomass and coal can be considered as potential fuels for gasification and further syngas production and methanol synthesis (Takezawa *et al.* 1987). Adding sufficient hydrogen to the syngas to convert all of the biomass into methanol produces double the amount of methanol produced from the same biomass base. Waste material can be partially converted to methanol, for which the product yield for the conversion process is estimated to be 185 kg of methanol per metric ton of solid waste (Sorenson 1983). Agricultural methanol/ethanol is at present more expensive than ethanol synthesized from ethylene and methanol produced from natural gas (Grassi 1999).

On the basis of a comparison of the revenues and costs of different conventional technologies, methanol synthesis from natural gas reforming can be chosen as the downstream process. This technology can be used to reduce the cost of gas transportation, helps satisfy the demand for cleaner fuels, and fetches higher revenues. The reactions in the methane reforming process are:

Steam reforming:

$$CH_4 + H_2O \rightarrow CO + 3H_2 \tag{3.17}$$

Partial oxidation:

$$CH_4 + 1/2O_2 \rightarrow CO + 2H_2 \tag{3.18}$$

Table 3.10 Yields of products from the steam reforming of methane over Ni/Ce–ZrO_2 plus Ni/YSZ at different methane to water ratios at 1,175 K

CH_4 to H_2O ratio	Hydrogen (%)	Carbon monoxide (%)	Carbon dioxide (%)
1:3	76.1	17.5	6.5
1:4	79.0	14.0	7.0
1:5	82.8	10.1	7.1

Carbon dioxide reforming:

$$CH_4 + CO_2 \rightarrow 2CO + 2H_2 \tag{3.19}$$

The biggest advantage of this process is the replacement of the highly endothermic steam reforming process by the exothermic partial oxidation process. For the partial oxidation process, the fuel is introduced into the reactor with a controlled amount of oxygen. The oxidation reaction is highly exothermic and the reaction rates are generally very rapid. For methanol synthesis, the ideal H_2 to CO ratio is 2:1.

The catalyst chosen is Ni (2.5 wt%) on an Al_2O_3 support. Nickel shows good activity for hydrogen transfer and is commonly used in stationary plants for hydrogen production by both partial oxidation and steam reforming. The reforming reactor is a packed-bed type. The reaction conditions are 775 K, 1 atm, and 0.1-s residence time.

The economic analysis examines the cost of each process according to the natural gas throughput and product rates. The capital investment required for a conventional methanol production facility of approximately 100 metric tons per day is $10 million. The major factors affecting this capital cost include the costs for steam production, methanol synthesis, distillation, utilities, the oxygen unit, and glycol separation.

3.5 Summary

Gas hydrates are icelike crystalline solids that form from mixtures of water and light natural gases such as methane, carbon dioxide, ethane, propane, and butane. They occur where pressure, temperature, gas saturation, and local chemical conditions combine to make them stable. Methane is the dominant component among other hydrocarbon gases in sediments.

Gas hydrates were discovered in 1810 by Sir Humphrey Davy, and were considered to be a laboratory curiosity. In the 1930s, clathrate formation turned out to be a major problem, clogging pipelines during transportation of gas under cold conditions.

Subseabed methane within the continental margin sediments is produced primarily by microbial or thermogenic processes. In the microbial process, organic debris is decomposed by a complex sequence (methanogenesis) into methane by bacteria in an anoxic environment. Organic matter is composed of carbon, hydrogen, and phosphorus in the ratio of 106:16:1, and decomposition results in production of methane.

Gas hydrate consists of three general structure types. Depending on the size of the guest molecule, natural gas hydrates can consist of any combination of three crystal structures: (1) sI, (2) sII, and (3) sH. When pure liquid water freezes, it crystallizes with hexagonal symmetry, but when it "freezes" as a hydrocarbon hydrate, it does so with cubic symmetry for sI and sII, reverting to hexagonal symmetry for sH.

References

Achenbach E, Riensche E (1994) Methane/steam reforming kinetics for solid oxide fuel cells. J Power Sources 52:283

Andersland OB, Ladanyi B (1994) Frozen ground engineering. Chapman and Hall, New York

Bradford MCJ, Vannice MA (1999) CO_2 reforming of CH_4. Catal Rev Sci Eng 41:1–42

Brewer PG, Orr FM, Friederich G, Kvenvolden KA, Orange DL, McFarlane J, Kirkwood W (1997) Deep ocean field test of methane hydrate formation from a remotely operated vehicle. Geology 25:407–410

Buchanan P, Soper AK, Thompson H, Westacott RE, Creek JL, Hobson G, Koh CA (2005) Search for memory effects in methane hydrate: structure of water before hydrate formation and after hydrate decomposition. J Chem Phys 123:164507

Buffett B, Archer D (2004) Global inventory of methane clathrate: sensitivity to changes in the deep ocean. Earth Planet Sci Lett 227:185–199

Buffett BA, Zatsepina OY (2000) Formation of gas hydrate from dissolved gas in natural porous media. Mar Geol164:69–77

Choudhary TV, Goodman DW (2000) Methane activation on Ni and Ru model catalysts. J Mol Catal A 163:9–18

Collett TS (2002) Energy resource potential of natural gas hydrates. AAPG Bull 86:1971–1992

Dallimore SR, Collett TS (1999) Regional gas hydrate occurrences, permafrost conditions, and Cenozoic geology, Mackenzie Delta area. In: Collett TS (ed) Scientific results from JAPEX/JNOC/GSC Mallik 2L-38 gas hydrate research well, Mackenzie Delta, Northwest Territories, Canada. Geological Survey of Canada, Ottawa, pp 31–43

Desa E (2001) Submarine methane hydrates potential fuel resource of the 21st century. Proc AP Akad Sci 5:101–114

Dickens GR (2003) Rethinking the global carbon cycle with a large, dynamic and microbially mediated gas hydrate capacitor. Earth Planet Sci Lett 213:169–183

Dickens GR, Castillo MM, Walker JCG (1997) A blast of gas in the latest Paleocene: simulating first order effects of massive dissociation of oceanic methane hydrates. Geology 25:259–262

Dicks AL, Pointon KD, Siddle A (2000) Intrinsic reaction kinetics of methane steam reforming on a nickel/zirconia anode. J Power Sources 86:523

Durham WB, Kirby SH, Stern LA, Zhang W (2003) The strength and rheology of methane clathrate hydrate. J Geophys Res 108(B4):2182

Edmonds B, Moorwood R, Szczepanski R (1996) A practical model for the effect of salinity on gas hydrate formation. SPE paper 35569. Paper presented at the Society of Petroleum Engineers European production operations conference and exhibition, Stavanger, 16–17 April 1996

European Renewable Energy Council (2006) Renewable energy scenario by 2040. European Renewable Energy Council, Brussels

Farkhondeh M, Gheisi AR (2002) An introduction to natural gas hydrate. Transportation. Methane gas hydrate report. Tehran University

Gao S, House W, Chapman WG (2005) NMR/MRI study of clathrate hydrate mechanisms. J Phys Chem B 109:19090–19093

Goldberg D, Saito S (1998) Detection of gas hydrates using downhole logs. In: Henriet J.-P, Mienert J (eds) Gas hydrates – relevance to world margin stability and climate change. Geological Society, London, pp 129–132

Grassi G (1999) Modern bioenergy in the European Union. Renew Energy 16:985–990

Guerin G, Goldberg D, Meltse A (1999) Characterization of in situ elastic properties of gas hydrate-bearing sediments on the Blake Ridge. J Geophys Res 104:1778–17795

Gupta AK (2004) Marine gas hydrates: their economic and environmental importance. Curr Sci 86:1198–1199

Hacisalihoglu B, Demirbas AH, Hacisalihoglu S (2008) Hydrogen from gas hydrate and hydrogen sulfide in the Black Sea. Energy Educ Sci Technol 21:109–115

Handa YP, Stupin D (1992) Thermodynamic properties and dissociation characteristics of methane and propane hydrates in 70 A radius silica gel pores. J Phys Chem 96:8599–8603

Handy MR (1994) Flow laws for rocks containing two non-linear viscous phases: a phenomenological approach. J Struct Geol 16:287–301

Helgerud MB (2001) Wave speeds in gas hydrate and sediments containing gas hydrate: a laboratory and modeling study. PhD thesis, Stanford University Press, Stanford

Hiraia H, Tanaka T, Kawamura T, Yamamoto Y, Yagi T (2004) Structural changes in gas hydrates and existence of a filled ice structure of methane hydrate above 40 GPa. J Phys Chem Solids 65:1555–1559

Holbrook WS, Hoskins H, Wood WT, Stephen RA, Lizarralde D (1996) Methane hydrate and free gas on the Blake Ridge from vertical seismic profiling. Science 273:1840–1843

Hornbach MJ, Holbrook WS, Gorman AR, Hackwith KL, Lizarralde D, Pecher I (2003) Direct seismic detection of methane hydrate on the Blake Ridge: Geophysics 68:92–100

Hovland M, Judd AG (eds) (1998) Seabed pockmarks and seepages: impact on geology, biology and the marine environment. Graham and Trotman, London

Hu YH, Ruckenstein E (2006) Clathrate hydrogen hydrate – a promising material for hydrogen storage. Angew Chem 45:2011–2013

Iwasa N, Kudo S, Takahashi H, Masuda S, Takezawa N (1993) Highly selective supported Pd catalysts for steam reforming of methanol. Catal Lett 19:211–216

Jenkins BM, Baxter LL, Miles TR Jr, Miles TR (1998) Combustion properties of biomass. Fuel Process Technol 54:17–46

Katsumasa K, Koga K, Tanaka H (2007) On the thermodynamic stability of hydrogen clathrate hydrates. J Chem. Phys 127:044509

Koh CA (2002) Towards a fundamental understanding of natural gas hydrates. Chem Soc Rev 31:157–167

Koo KY, Roh H-S, Seo YT, Seo DJ, Yoon WL, Park SB (2008a) Coke study on MgO-promoted Ni/Al_2O_3 catalyst in combined H_2O and CO_2 reforming of methane for gas to liquid (GTL) process. Appl Catal A Gen 340:183–190

Koo KY, Roh H-S, Seo YT, Seo DJ, Yoon WL, Park SB (2008b) A highly effective and stable nano-sized Ni/MgO–Al_2O_3 catalyst for gas to liquids (GTL) process. Int J Hydrogen Energy 33:2036–2043

Kvenvolden K (1993a). Gas hydrates – geological perspective and global change. Rev Geophys 31:173–187

Kvenvolden KA (1993b) Gas hydrates as a potential energy resource – a review of their methane content. In: Howell DG (ed) The future of energy gases – US Geological Survey professional paper 1570. United States Government Printing Office, Washington, pp 555–561

Kvenvolden KA (1998) A primer on the geological occurrence of gas hydrate. Geol Soc Lond Spec Publ 137:9–30

Kvenvolden KA, Lorenson TD (2000) The global occurrence of natural gas hydrate. In: Paull CK, Dillon WP (eds) Natural gas hydrates: occurrence, distribution, and dynamics. Geophysical monograph 124. American Geophysical Union, Washington, pp 3–18

Laosiripojana N, Assabumrungrat S (2007) Catalytic steam reforming of methane, methanol, and ethanol over Ni/YSZ: the possible use of these fuels in internal reforming SOFC. J Power Sources 163:943–951

Lerche I, Bagirov E (1998) Guide to gas hydrate stability in various geological settings. Mar Pet Geol 15:427–438

Mahajan D, Taylor CE, Mansoori GA (2007) An introduction to natural gas hydrate/clathrate: the major organic carbon reserve of the Earth. J Pet Sci Eng 56:1–8

Mao WL, Mao HK, Goncharov AF, Struzhkin VV, Guo Q, Hu J. Shu J, Hemley RJ, Somayazulu M, Zhao Y (2002) Hydrogen clusters in clathrate hydrate. Science 297:2247–2249

Martinson O (1994) Mass movements. In: Maltman A (ed) Geological deformation of sediments. Chapman and Hall, New York, pp 127–165

Matsumoto R, Watanabe Y, Satoh M, Okada H, Hiroki Y, Kawasaki M, ODP Leg 164 Shipboard Scientific Party (1996) Distribution and occurrence of marine gas hydrates – preliminary results of ODP Leg 164: Blake Ridge Drilling. J Geol Soc Jpn 102:932–944

Mazzini A, Ivanov MK, Parnell J, Stadnitskaia A, Cronin BT, Poludetkina E, Mazurenko L, van Weering TCE (2004) Methane-related authigenic carbonates from the Black Sea: geochemical characterization and relation to seeping fluids. Mar Geol 212:153–181

McCarthy VN, Jordan KD (2006) Structure and stability of the $(H_2O)_{21}$ and $(H_2O)_{20}\cdot(H_2S)$ clusters: relevance of cluster systems to gas hydrate formation. Chem Phys Lett 429:166–168

Minh NQ, Takahashi T (1995) Science and technology of ceramic fuel cells. Elsevier, Amsterdam

Morrison RT, Boyd RN (1983) Organic chemistry, 4th edn. Allyn and Bacon, Singapore

Mulder T, Alexander J (2000) The physical character of subaqueous sedimentary density flows and their deposits. Sedimentology 48:269–299

Ohmura R, Shigetomi T, Mori YH (1999) Formation, growth and dissociation of clathrate hydrate crystals in liquid water in contact with a hydrophobic hydrate-forming liquid. J Cryst Growth 196:164–173

Patchkovskii S, Tse JS (2003) Thermodynamic stability of hydrogen clathrates. Proc Natl Acad Sci USA 100:14645–14650

Peña MA, Gómez JP, Fierro JLG (1996) New catalytic routes for syngas and hydrogen production. Appl Catal A 144:7–57

Popescu I, De Batist M, Lericolais G, Nouzé H, Poort J, Panin N, Versteeg W, Gillet H (2006) Multiple bottom-simulating reflections in the Black Sea: potential proxies of past climate conditions. Mar Geol 227:163–166

Roh H-S, Jun K-W, Baek S-C, Park S-E (2002) Carbon dioxide reforming of methane over Ni/q-Al_2O_3 catalysts: effect of Ni content. Bull Korean Chem Soc 23:1166–1168

Rostrup-Nielsen JR (1984) Catalytic steam reforming. In: Anderson JR, Boudart M (eds) Catalysis science and technology. Springer, Berlin

Sanloup C, Mao H, Hemley RJ (2002) High-pressure transformations in xenon hydrates. Proc Natl Acad Sci USA 99:25–28

Satoh M, Maekawa T, Okuda Y (1996) Estimation of amount of methane and resources of gas hydrates in the world and around Japan. J Geol Soc Jpn 102:959–971

Schicks JM, Erzinger J, Ziemann MA (2005) Raman spectra of gas hydrates – differences and analogies to ice Ih and (gas saturated) water. Spectrochim Acta A Mol Biomol Spectrosc 61:2399–2403

Schuth F (2005) Hydrogen and hydrates. Nature 434:712–713

Shipley TH, Houston MH, Buffler RT, Shaub FJ, McMillen KJ, Ladd JW, Worzel JL (1979) Seismic evidence for widespread possible gas hydrate horizons on continental slopes and rises. AAPG Bull 63:2204–2213

Sloan ED (1990) Natural gas hydrate phase equilibria and kinetics; understanding the state-of-the-art. Rev Inst Fr Pet 45:245-266

Sloan ED Jr (1998) Clathrate hydrates of natural gases, 2nd edn. Dekker, New York

Sloan ED, Koh CA (2008) Clathrate hydrates of natural gases, 3rd edn. CRC, Boca Raton

Sorensen HA (1983) Energy conversion systems. Wiley, New York

Stoll RD, Bryan GM (1979) Physical properties of sediments containing gas hydrate. J Geophys Res 84(B4):1629–1634

Suess E, Bohrmann G, Rickert D, Kuhs WF, Torres ME, Thehu A, Linke P (2002) Properties and fabric or near-surface methane hydrates at Hydrate Ridge, Cassadia Margin. In: Proceedings of the fourth international conference on gas hydrates, Yokohama, 29–23 May

Takezawa N, Shimokawabe M, Hiramatsu H, Sugiura H, Asakawa T, Kobayashi H (1987) Steam reforming of methanol over Cu/ZrO_2. Role of ZrO_2 support. React Kinet Catal Lett 33:191–196

Tohidi B, Anderson R, Clennell MB, Burgass RW, Biderkab AB (2001) Visual observation of gas-hydrate formation and dissociation in synthetic porous media by means of glass micromodels. Geology 29:867–870

Torres ME, Wallmann K, Trhu AM, Bohrmann G, Borowski WS, Tomaru H (2004) Gas hydrate growth, methane transport, and chloride enrichment at the southern summit of Hydrate Ridge, Cascadia margin off Oregon. Earth Planet Sci Lett 226:225–241

Tucholke BE, Bryan GM, Ewing JI (1977) Gas-hydrate horizons detected in seismic-profiler data from the western North Atlantic. AAPG Bull 61:698–707

Tullis TE, Horowitz FG, Tullis J (1991) Flow laws of polyphase aggregates from end-member flow laws. J Geophys Res 96:8081–8096

UEIC (1989) Hydrogen sulfide. In: Ullmann's encyclopedia of industrial chemistry, 5th edn, vol A13. VCH, Weinheim, pp 467–485

Waite WF, Winters WJ, Mason DH (2004) Methane hydrate formation in partially water-saturated Ottawa sand. Am Mineral 89:1202–1207

Winters WJ, Pecher IA, Waite FW, Mason DH (2004) Physical properties and rock physics models of sediment containing natural and laboratory-formed methane gas hydrate. Am Mineral 89:1221–1227

Yentekakis IV, Neophytides SG, Kaloyiannis AC, Vayenas CG (1993) A solid oxide fuel cell using an exothermic reaction as the heat source. Proceeding of the 3rd international symposium on solid oxide fuel cells, vol 4, Honolulu

Zhang W, Durham WB, Stern A, Kirby SH (1999) Experimental deformation of methane hydrate: new results. Eos Trans AGU 80(17):S337

[illegible]

Chapter 4
Methane Gas Hydrate: as a Natural Gas Source

4.1 Introduction

Natural gas hydrates are members of a highly varied class of substances called clathrates. Gas hydrates are icelike substances composed of a host lattice of water molecules and one or more of a potential suite of guest molecules which at normal temperatures and pressures occur in the gaseous phase and are capable of physically fitting into the interstices of the water-ice lattice. This suite includes the noble gases (the elements helium, neon, krypton, argon, xenon, and radon), the halogens chlorine, bromine, iodine, and astatine, and hydrogen sulfide, sulfur trioxide, sulfur hexafluoride, and carbon dioxide. Significantly, it also includes the low molecular weight hydrocarbons methane, ethane, propane, and the pentanes.

In the last few years, many governments (including those of the USA, Canada, Russia, India, and Japan) have become very interested in the possibilities of methane hydrates. Gas hydrates represent one of the world's largest untapped reservoirs of energy and, according to some estimates, have the potential to meet global energy needs for the next 1,000 years.

Considerable knowledge has been developed about their occurrence in nature and their properties. However, the issues of the technology and the economics of natural gas production from gas hydrates are less well understood. Gas hydrates are increasingly recognized as a potential future energy resource, based on the vast amounts of methane trapped within them (Kvenvolden 1993). They are also of considerable interest because of their potential role in global climate change. The amount of organic carbon in gas hydrates is estimated to be twice that in all other fossil fuels combined. From an energy resource point of view, the enormous amounts of gas hydrate under the ocean and beneath Arctic permafrost represent an estimated 53% of all fossil fuel (coal, oil, and natural gas) reserves on Earth, about 10,000 gigatons (Lee and Holder 2001). The difficulty in recovering this source of energy is that the fuel is in solid form and is not amenable to conventional gas and oil recovery techniques (Lee and Holder 2001).

A. Demirbas, *Methane Gas Hydrate*,

Proposed methods of gas recovery from hydrates generally deal with dissociating or melting *in situ* gas hydrates by heating the reservoir beyond the temperature of hydrate formation, or decreasing the reservoir pressure below hydrate equilibrium. Models have been developed to evaluate natural gas production from hydrates by both depressurization and heating methods.

Methane gas hydrates are of interest primarily for three reasons:

1. Gas from hydrate may be a new clean energy source. It is now recognized that there are huge amounts of natural gas, mainly methane, tied up in gas hydrates globally. Methane gas hydrates are a potential energy resource.
2. Natural gas hydrate may play a role in climate change. Methane is a strong greenhouse gas, so its escape to the atmosphere from natural gas hydrate could result in global warming.
3. There are important production problems. Gas hydrate is a hazard in conventional hydrocarbon exploration, from shallow gas release and from seafloor instability, especially in the Arctic and in deep water where hydrate is stable.

Considering the planet as a whole, the quantity of natural gas in sedimentary gas hydrates greatly exceeds the conventional natural gas resources (Kvenvolden 1993). As a result, numerous studies have discussed the energy resource potential of gas hydrates (Collett 2000, 2002; Buffett and Zatsepina 2000; Kvenvolden 1993; Brewer *et al.* 1997). However, utilization of gas hydrates as an energy resource has been largely inhibited by the lack of economic methods for production for most hydrate accumulations, especially marine shelf hydrates. There are three methods to obtain methane from gas hydrates: (1) the depressurization method; (2) the thermal stimulation method, and (3) the chemical inhibition method. The thermal stimulation method is quite expensive. The chemical inhibitor injection method is also expensive. The depressurization method may prove more useful than the other production methods.

4.2 Importance of Methane Hydrates

Methane can be produced from natural gas hydrates, or methane hydrates. The presence of gas hydrates in oceanic sediments was first postulated on the basis of seismic observations. Gas hydrates represent one of the world's largest untapped reservoirs of energy and, according to some estimates, have the potential to meet global energy needs for the next 1,000 years. Gas hydrates are potentially one of the most important energy resources for the future. Methane gas hydrates are increasingly considered to be a potential energy resource (Collett 2002). Gas hydrates are of considerable interest because of their potential as an energy resource and because of their potential role in global climate change. From an energy resource point of view, the enormous amounts of gas hydrate under the ocean and beneath Arctic permafrost represent an estimated more than 50% of all carbonaceous fuel reserves on Earth (Kvenvolden 1999).

Gas hydrates are crystalline solid compounds, consisting of a gas molecule surrounded by a cage of water molecules. They are stable under the high pressure and low temperature typical of deepwater sediments of the world ocean. Gas hydrate, or methane hydrate, is composed of natural gas molecules trapped inside ice. Gas hydrates contain twice as much carbon as all other fossil fuels (coal, natural gas, and oil) combined. The amount of methane trapped in marine sediments as a hydrate represents such an immense carbon reservoir that it must be considered a dominant factor in estimating unconventional gas energy resources. Significant safety and environmental concerns are also associated with the presence of natural gas hydrates, ranging from their possible impact on the safety of conventional drilling operations to the influence on the earth's periodic natural releases into the atmosphere of large volumes of hydrate-sourced methane or its derivative carbon dioxide.

What are the energy sources of the future? Some scientists believe methane hydrates are one of the answers. On Earth, their global abundance and distribution suggest that they may become energy resources of the future (Collett 2000; Kvenvolden 1993). With increasing energy demand and depleting energy resources, gas hydrates may serve as a potentially important resource for future energy requirements. The conditions suitable for the occurrence of gas hydrate exist in a few hundred meters of the rapidly accumulating continental margin organic-rich sediments (Gupta 2004).

Hydrocarbon clathrate hydrates are important intermolecular compounds and naturally occurring minerals that occur on submarine continental margins and regions of Arctic permafrost (Kvenvolden 1993, 2000; Sloan 1998b; Durham *et al.* 2003). They are also expected to occur within medium-sized to large icy moons of the outer solar system (Loveday *et al.* 2001; Lunine and Stevenson 1985), and in the polar regions of Mars (Miller and Smythe 1970; Jakosky *et al.* 1995).

It has become increasingly evident that naturally occurring gas hydrates are important components of the shallow geosphere and are of societal concern in at least three major ways: resource, hazard, and climate (Kvenvolden 1993). There are two reasons why gas hydrates are attractive as a potential resource. The first one is the enormous amount of methane that is apparently sequestered within clathrate structures at shallow sediment depths within 2,000 m of the earth's surface. The second one is the wide geographic distribution of the gas hydrates. It was mentioned that the energy potential of methane hydrates is considerably greater than that of the other unconventional sources of gas, such as coal beds, tight sands, black shales, deep aquifers, and conventional natural gas (MacDonald 1990a). The resource potential of marine gas hydrate is yet to be ascertained, but considering the possibility of enormous gas reservoirs, gas hydrates will continue to attract attention until their development potential is measured (Gralus 2001).

Methane hydrates are considered to be a major potential source of hydrocarbon energy and could be important in meeting natural gas demand in the future. The hydrates and any free gas trapped below the hydrate stability zone may provide a significant hydrocarbon resource in the future (Kvenvolden and McMenamin 1980; Kvenvolden 1993).

4.3 Global Climate Change

Today's world is facing two environmental problems: global warming and air pollution. Both of these are linked to the heavy use of fossil fuels. Global consumption of fossil fuels is staggering and is increasing at an alarming rate. The quadrupling of oil prices in the 1970s and the rapid increase in recent years, the growing awareness of energy-related pollution, and the possibility of climate change have all contributed to a reevaluation of energy use. The result has been an improvement in the efficiency with which energy is used in industry and power generation as well as in lighting, household appliances, transportation, and heating and cooling of buildings. This more efficient use of energy is a major factor contributing to the improvements in energy utilization that have occurred historically in almost all OECD countries, and more recently in many transition economies, as well as in some in fast-growing developing countries. Increasingly, engineering practitioners and managers need to know how to respond to the challenges of integrating environmentally conscious technologies, techniques, strategies, and objectives into their daily work, and, thereby, find opportunities to lower costs and increase profits while managing to limit environmental impacts. Engineering practitioners and managers also increasingly face challenges in complying with changing environmental laws.

Reducing the use of fossil fuels would considerably reduce the amount of carbon dioxide produced, as well as reducing the levels of the pollutants which cause acid rain. This can be achieved by either using less fossil energy altogether or using alternative energy sources.

Large reductions in emissions of carbon dioxide to the atmosphere may be needed to avoid climate change. Capture and storage of carbon dioxide, in combination with other carbon dioxide abatement techniques, could enable these large reductions to be achieved. Storage of carbon dioxide in geological structures could have a beneficial effect on the oil and gas industries in terms of enhanced oil and gas production. The Kyoto Protocol and its flexible mechanisms will provide opportunities to develop projects in developing countries and generate carbon credits.

4.3.1 Air Pollution

Air pollution is a mixture of natural and man-made substances in the air we breathe, such as fine particles produced by the burning of fossil fuels, ground-level ozone, which is a reactive form of oxygen that is a primary component of urban smog, and noxious gases such as sulfur dioxide, nitrogen oxides, carbon monoxide, and chemical vapors. The health effects of air pollution have been reported in research studies over the past 30 years. These effects include respiratory diseases such as asthma, cardiovascular diseases, changes in lung function, and death. There is mounting evidence that exposure to air pollution has long-term effects on lung development in children.

Air pollution can cause health problems and it can also damage the environment and property. It has caused thinning of the protective ozone layer of the atmosphere, which is leading to climate change.

Air quality standard (AQS) is the prescribed level of a pollutant in the outside air that should not be exceeded during a specific time period to protect public health. Air pollution is the presence of polluting gases and suspended particles in the atmosphere in excess of the AQS.

Air quality criteria are the varying amounts of pollution and lengths of exposure at which specific adverse effects to health and comfort take place. The main air pollutants are carbon monoxide, lead, nitrogen dioxide, ozone, persistent organic pollutants, suspended particulate matter, and sulfur dioxide (Demirbas 2005).

Coal and oil contain various organic sulfur compounds, such as benzothiophenes, dibenzothiophenes, benzonaphthothiophenes, penanthrothiophenes, and many other sulfur-containing compounds with up to seven aromatic rings.

The burning of coal, oil, and natural gas, as well as deforestation and various agricultural and industrial practices are altering the composition of the atmosphere and contributing to climate change. These human activities have led to increased atmospheric concentrations of a number of greenhouse gases, including carbon dioxide, methane, nitrous oxide, chlorofluorocarbons (CFCs), and ozone in the lower part of the atmosphere.

Coal conversion generates a number of gaseous and solid pollutants:

1. Nitrogen oxides
2. Sulfur oxides
3. Carbon dioxide
4. Fine particulates
5. Heavy metals

When exposed to high temperatures, nitrogen in the air and the fuel can be oxidized, forming a number of different compounds, such as nitric oxide (NO), nitrogen dioxide (NO_2), and nitrous oxide (N_2O). Nitrogen oxides (NO_x), which include NO and NO_2, are responsible for the formation of acid rain, photochemical smog, and ground-level ozone. The formation of NO and NO_2 is favored at elevated temperatures (about 1,500 K) as found in pulverized coal flames, whereas the formation of N_2O is favored at lower temperatures (about 1,100 K) like those found in fluidized bed combustors. The nitrogen in fuel is oxidized at all combustion temperatures, but that in air is oxidized generally above 1,300 K (Kutz 2007).

Sulfur dioxide (SO_2) is formed when the sulfur in coal, either in pyretic or in organically bonded form, is oxidized. SO_2 is the leading contributor to acid rain formation and is an irritant to the lungs. Coal combustion makes up the majority of SO_2 emissions from utilities, which in turn account for about one third of the total SO_2 emissions from all sources.

In a coal-burning plant, a portion of the ash is released to the atmosphere through the stack, with the very fine particulates having harmful health effects. Particulate matter is an air pollutant consisting of a mixture of particles that can be solid, liquid, or both, are suspended in the air, and represent a complex mixture of organic and inorganic substances. Particulate matter smaller than about 10 μm can settle in the bronchial tubes and lungs, particles smaller than 2.5 μm can penetrate directly into the lung, and particles smaller than 1 μm can penetrate into the alveolar region of the lung and tend to be the most hazardous when inhaled. Fine particulates can have significant harmful health effects if exposure occurs over extended periods.

Most heavy metals such as lead, tin, and magnesium are collected in particulate control systems, but owing to its low vaporization point (629 K), much of the mercury found in coal escapes into the environment, making coal combustion the primary source for mercury emissions. Mercury vapor in the flue gas precipitates out into the environment and bioaccumulates in organisms such as fish, where it is often transformed into methylmercury, a highly toxic organic compound. Fish species that are high up in the food chain contain high concentrations of mercury, because they eat many smaller fish that have small amounts of mercury in them. The US Food and Drug Administration has an action level for methylmercury in commercial marine and freshwater fish of 1.0 ppm, and in Canada the limit for the total mercury content is 0.5 ppm (Kutz 2007).

Mercury is highly toxic and persistent. It bioaccumulates in marine organisms and can be transformed into methylmercury by the action of bacteria. Methylmercury is an organic form which is, in turn, more toxic and more readily bioaccumulated into internal organs, particularly the liver and muscle tissues. Mercury results from both natural and anthropogenic release into the environment, the bulk of which comes from the combustion of fossil fuels and waste incineration. Industrial processes and manufactured goods such as thermometers, dental fillings, and fluorescent lights also make a significant contribution.

High levels of mercury may present a significant contributory factor to cetacean mortality, particularly in animals weakened by disease and therefore less able to detoxify organic mercury as efficiently as healthy animals. Methylmercury exposure in humans can cause irreversible neurological damage. Symptoms can include impaired vision, speech, and hearing, loss of coordination, reproductive disorders, paralysis, and cerebral palsy. Severe cases may result in coma or death.

Lead is present in petrol, diesel, lead batteries, paints, hair dye products, etc. Lead affects children in particular. It can cause nervous system damage and digestive problems and, in some cases, cancer. Recently, there has been growth in public concern about the addition of lead to petrol. Apart from octane improvement, trace lead is also a valve lubricant. Lead is not the only environmental problem with petrol which has been a cause of public concern (Kalač *et al.* 2004). The average AQS value of lead in respirable air is 0.75–1.0 μg/m^3. When released to the environment, lead has a long residence time compared with most pollutants, remaining accessible to the food chain and to human metabolism far into the future (Sauve *et al.* 1997). The toxic effects of lead include damage to the kidneys and the cardiovascular and nervous system. Of particular concern is the effect of relatively low exposure on cognitive and behavioral development in children (Pirkle *et al.* 1998).

Research carried out in recent years has proved that besides CO_2, CO, SO_x, NO_x, and particulate materials being emitted in significant amounts, emissions are also responsible for significant amounts of persistent organic pollutants, including polycyclic aromatic hydrocarbons, polychlorinated dibenzo-*p*-dioxins, polychlorinated dibenzofurans, and volatile organic compounds such as aliphatic hydrocarbons, benzene and its derivatives (benzene, toluene, and xylene), aldehydes, and ketones, and also phenol and its alkyl derivatives, heterocompounds of nitrogen, sulfur etc., and heavy metals (Kubica 1997). The combustion or incineration of various wastes or natural materials containing chlorine can lead to the formation and emission of polycyclic aromatic hydrocarbons, polychlorinated dibenzo-*p*-dioxins, polychlorinated dibenzofurans, chlorohydrocarbons, and other species. (Chagger *et al.* 1998).

Dioxin and furans are a class of 205 halogenated aromatic hydrocarbons which can have detrimental impacts on human health at very low concentrations (Kannan *et al.* 2007). Dioxin forms when organic substances are heated in the presence of chlorine. Dioxin in sewage sludge may come from industrial discharge to the system, or from airborne dioxin settling onto land areas, and then being washed into storm sewers, and then into the treatment plant. The dioxin is not formed in the sewage treatment process (Tame *et al.* 2007).

Dioxins, formed in any combustion process where carbon, oxygen, and chlorine are present, are a subject of major interest owing to their carcinogenicity. Much research has been carried out to study emissions from hazardous and municipal waste incinerators (Lavric *et al.* 2004).

About 98% of carbon emissions result from fossil fuel (coal, oil, and natural gas) combustion. Reducing the use of fossil fuels would considerably reduce the amount of carbon dioxide produced, as well as reducing the levels of the pollutants. Indeed, much of the variation in the estimates of the cost to control carbon emissions revolves around the availability and cost of carbon-free technologies and carbon-reducing technologies, such as energy-efficiency and energy-conservation equipment. This can be achieved by either using less energy altogether or using alternative energy sources. Much of the current effort to control such emissions focuses on advancing technologies that emit less carbon (e.g., high-efficiency combustion) or no carbon, such as nuclear, hydrogen, solar, wind, geothermal, and other renewable energy sources, or on using energy more efficiently, and on developing innovative technologies and strategies to capture and dispose of carbon dioxide emitted during fossil fuel combustion. In addition, overall carbon dioxide emissions can be reduced by biomass combustion because biomass is a carbon dioxide neutral fuel.

$$\begin{array}{llll} CH_4 & + 2O_2 & \rightarrow CO_2 & + 2H_2O \\ 1.00\,g & & \;\;\;\; 2.75\,g & \end{array} \tag{4.1}$$

$$\begin{array}{llll} 2C_4H_{10} & + 13O_2 & \rightarrow 8CO_2 & + 10H_2O \\ 1.00\,g & & \;\;\;\; 3.03\,g & \end{array} \tag{4.2}$$

$$\begin{array}{lll} C & + O_2 & \rightarrow CO_2 \\ 1.00\,g & & \;\;\;\; 3.67\,g \end{array} \tag{4.3}$$

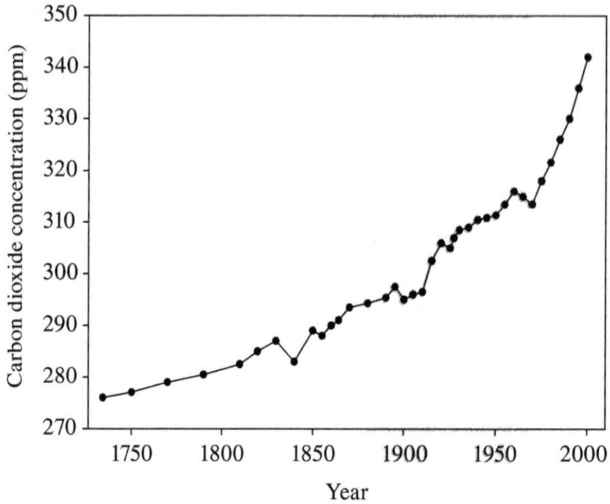

Fig. 4.1 Carbon dioxide concentration by year

Among the fossil fuels, natural gas is the least responsible for carbon dioxide emissions. LPG causes more carbon dioxide than does natural gas.

Presently, coal is responsible for 30–40% of world carbon dioxide emissions from fossil fuels. About 98% of carbon emissions result from fossil fuel (coal, oil, and natural gas) combustion. Figure 4.1 shows a plot of carbon dioxide concentrations by year. Global warming has been increasingly associated with the contribution of carbon dioxide. Figure 4.2 shows the world carbon dioxide emissions between 1000 and 2000.

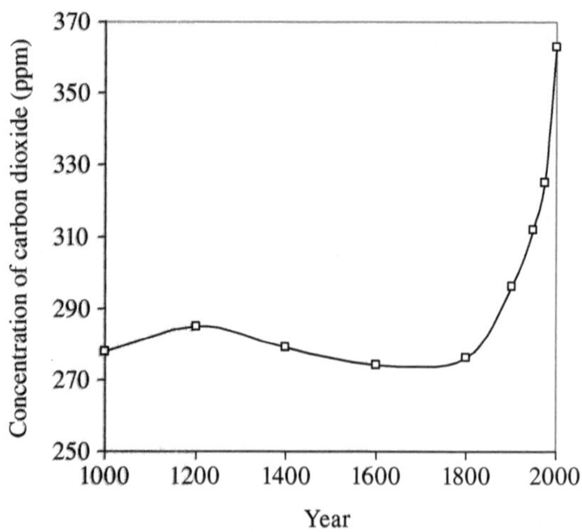

Fig. 4.2 World carbon dioxide emissions between 1000 and 2000 (Jean-Baptiste and Ducroux 2002)

4.3.2 Greenhouse Effect

The molar heat capacities of N_2, O_2, CO_2, CH_4, and SO_2 gases are 2.49, 2.51, 3.40, 3.34, and 3.76 cal/K at 288 K and 1 atm, respectively. The gases (they consist of three or more atoms) with heat capacities higher than those of O_2 and N_2 cause the greenhouse effect. Figure 4.3 shows the contribution of each gas to the greenhouse effect. Currently, it is estimated that CO_2 contributes about 55% to the anthropogenic greenhouse effect.

Carbon dioxide is main greenhouse gas associated with global warning and is produced in all combustion processes involving fossil fuels as well in other industrial processes such as cement production and sweetening of natural gas. The carbon content of coal is very high (50–89%), so it produces much larger amounts of carbon dioxide (carbon intensity) than do other fuels. Coal-fired power plants burning millions of tons of coal are therefore considered a major source of greenhouse gases. Increased levels of greenhouse gases in the atmosphere should lead to warmer temperatures on the earth's surface.

The greenhouse effect is the rise in temperature that the earth experiences because certain gases, such as water vapor, CO_2, N_2O, CFCs, CH_4, and other trace gases [hydrofluorocarbons (HFCs), perfluorocarbons (PFCs), sulfur hexafluorane (SF6), and trifluoromethyl sulfur pentafluoride (SF_5CF_3)], in the atmosphere trap energy from the sun. Increased levels of greenhouse gases in the atmosphere should lead to warmer temperatures on the earth's surface.

The four most important variable greenhouse gases whose atmospheric concentrations can be influenced by human activities are CO_2, CH_4, N_2O, and CFCs. Historically, CO_2 has been the most important, but over the past several decades other gases have assumed increasing significance. Collectively, they are projected

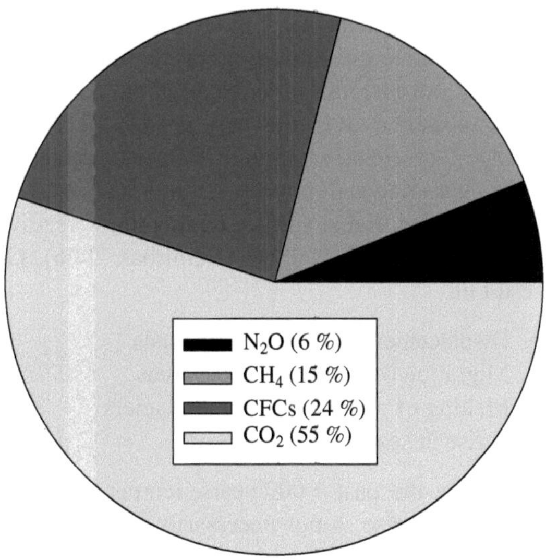

Fig. 4.3 Contribution of each gas to the greenhouse effect. *CFC* chlorofluorocarbons

to contribute, directly, about as much to potential global warming over the next 60 years as CO_2. Three trace gases, HFCs, PFCs, and SF6, are regulated under the 1997 Kyoto Protocol because of their global warming potential and because of the potential growth of their concentrations in the atmosphere. HFCs have been widely approved as substitutes for CFCs.

There are a large number of scientists who believe that human activities, which have increased atmospheric concentrations of CO_2 by more than one third over the past 100 years, may be leading to an increase in global average temperatures. However, this so-called global warming theory is not without challengers who argue that scientific proof supporting such theories is incomplete, and that many uncertainties remain surrounding the nature and direction of Earth's climate. Those changes have the potential to alter the way heat and energy are distributed among the land, ocean, atmosphere, and ice components of Earth's climate system. Ultimately, such changes might affect Earth's climate.

The development of advanced technology for biomass utilization is urgently required from the standpoint of global environmental protection measures such as CO_2 reduction and securing alternatives to fossil fuels. There is consensus among scientists that biomass fuels used in a sustainable manner result in no net increase in atmospheric CO_2. Some would even go as far as declaring that sustainable use of biomass will result in a net decrease in atmospheric CO_2. This is based on the assumption that all the CO_2 given off by the use of biomass fuels is taken in from the atmosphere by photosynthesis. Increased substitution of fossil fuels with biomass-based fuels would therefore help reduce the potential for global warming caused by increased atmospheric concentrations of CO_2.

4.3.3 Global Warming

Most scientists consider it likely that if the atmospheric concentrations of CO_2 and other so-called greenhouse gases continue to rise, the earth's climate will become warmer. An increase in average global temperatures of approximately 0.56 K has been measured over the past century. This increase is called "global climate change," or global warming. If current trends continue, the latest projections for future warming are between 1.5 and 5.8 K over the next century. Figure 4.4 shows the measured global surface temperatures relative to the average for the 130-year period from 1861 to 1990 (Demirbas 2006). Global warming of 1.5–5.8 K can result in:

- Displacement of agricultural zones
- Migration of tropical disease areas
- Melting of polar ice caps and glaciers
- A rise in sea level of 9–88 cm.

During the past 1,000 years, temperatures have naturally fluctuated by about 1 K, so this rise is not necessarily a result of the greenhouse effect. The latest

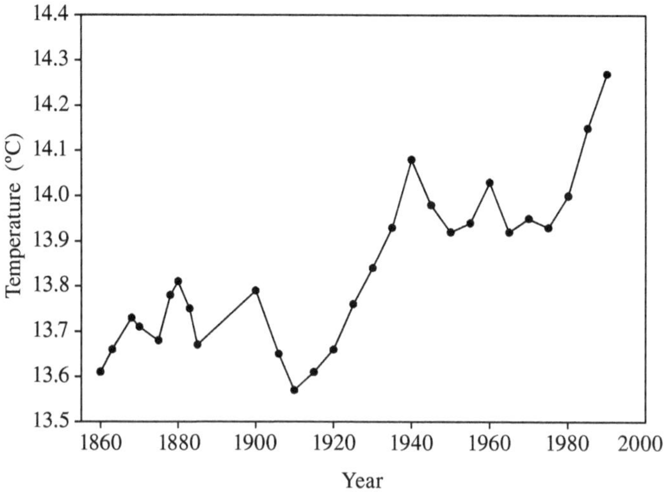

Fig. 4.4 Measured global surface temperatures relative to the average for the 130-year period from 1861 to 1990

computer models predict that global temperatures could rise by 1–3 K by the middle of this century if current trends persist. Global climate change could result in sea-level rises, changes to patterns of precipitation, increased variability in the weather, and a variety of other consequences. The production and human use of energy contributes 60% of the human impact on global climate change. In addition, the use of chemicals such as CFCs (15%), agriculture (12%), land-use modifications (9%), and other human activities (4%) also increase the level of greenhouse gases in the atmosphere. Presently, the rate of worldwide greenhouse gas emissions is increasing every year.

If we really wish for large metropolises, such as New York and London, not to be covered by the ocean waters one day in the near future, we should take the necessary precautions to decrease CO_2 emissions. There is a large potential for future increases in worldwide energy consumption driven by the increasing world population. If the coal industry is to participate in this growth, it is necessary to face the issue of CO_2 disposal.

Fossil-fuel-fired power plants account for approximately one third of the total CO_2 emissions. Because burning carbon must yield CO_2, it is impossible to modify fossil-fuel-fired power plants to stop producing CO_2. Therefore, if CO_2 emissions from fossil-fuel-fired power plants are to be reduced, the CO_2 produced must be captured and stored. Proposed methods to sequester CO_2 include terrestrial and ocean sequestration, geological sequestration, and mineral carbonation. The availability of a CO_2-fixation technology would serve as insurance in case global warming causes severe restrictions to be placed on CO_2 emissions. To prevent rapid climate change, it will be necessary to stabilize CO_2 as carbonate by the carbonation process.

The burning of coal, oil, and natural gas, as well as deforestation and various agricultural and industrial practices are altering the composition of the atmosphere and contributing to climate change. These human activities have led to increased atmospheric concentrations of a number of greenhouse gases, including CO_2, methane, nitrous oxide, CFCs, and ozone in the lower part of the atmosphere.

Replacing fossil fuels with wood has several distinct environmental implications. If wood is harvested at a rate which is sustainable, using it for energy purposes does not result in any net increase in atmospheric CO_2. This is because the CO_2 given off during the combustion of wood was recently taken from the atmosphere through photosynthesis. New tree growth will also soon reabsorb an equal amount. Released into the atmosphere by the burning of fossil fuels, CO_2 is the most important greenhouse gas responsible for global climate change. The use of wood as a source of energy can also reduce the amount of acid gases emitted into the atmosphere.

CO_2 and CH_4 (natural gas) are the most important of the greenhouse gases resulting from human activities. The use of fossil fuel currently accounts for 80–85% of the CO_2 being added to the atmosphere. Human activities have increased the concentration of CH_4 in the atmosphere by about 145% above what would be present naturally.

Nitrous oxide is produced by various agricultural and industrial practices. Human activities have increased the concentration of nitrous oxide in the atmosphere by about 15% above what would be present naturally.

CFCs have been used in refrigeration, air conditioning, and as solvents. However, the production of these gases is being eliminated under existing international agreements because they deplete the stratospheric ozone layer. Other fluorocarbons that are also greenhouse gases are being used as substitutes for CFCs in some applications, for example, in refrigeration and air conditioning. Although currently very small, their contributions to climate change are expected to rise.

The possibility of climate change raises many economic issues. The potential for such change introduces uncertainty in long-term decision-making and results in costs for the scientific investigation of the possibility, which may provide concomitant benefits, such as better understanding of the dynamics of climate or of biological responses to natural climate change.

As already mentioned, the four most important variable greenhouse gases, whose atmospheric concentrations can be influenced by human activities are CO_2, CH_4, N_2O, and CFCs. Historically, CO_2 has been the most important, but over the past several decades other gases have assumed increasing significance. Collectively, they are projected to contribute, directly, about as much to potential global warming over the next 60 years as CO_2. Three trace gases, HFCs, PFCs, and SF6, are regulated under the 1997 Kyoto Protocol because of their global warming potential and because of their potential growth of concentrations in the atmosphere. HFCs have been widely approved as substitutes for CFCs.

About 98% of carbon emissions result from fossil fuel (coal, oil, and natural gas) combustion. Reducing the use of fossil fuels would considerably reduce the amount of carbon dioxide produced, as well as reducing the levels of the pollutants. Indeed, much of the variation in the estimates of the cost to control carbon emissions revolves around the availability and cost of carbon-free technologies and carbon-reducing technologies, such as energy-efficiency and energy-conservation equipment. This can be achieved by either using less energy altogether or using alternative energy sources. Much of the current effort to control such emissions focuses on advancing technologies that emit less carbon (e.g., high-efficiency combustion) or no carbon, such as nuclear, hydrogen, solar, wind, geothermal, and other renewable energy sources, or on using energy more efficiently, and on developing innovative technologies and strategies to capture and dispose of carbon dioxide emitted during fossil fuel combustion. In addition, overall carbon dioxide emissions can be reduced by biomass combustion because biomass is a carbon dioxide neutral fuel.

The use of biomass fuels or bioenergy has the potential to greatly reduce our greenhouse gas emissions. Bioenergy generates about the same amount of carbon dioxide as fossil fuels, but every time a new plant grows, carbon dioxide is actually removed from the atmosphere. The net emission of carbon dioxide will be zero as long as plants continue to be replenished for bioenergy purposes. The energy crops, fast-growing trees, and grasses are called bioenergy sources. The use of bioenergy sources can also help increase profits for the agricultural industry.

4.3.4 Arguments Against Global Warming

There are some people who do not believe in global warming. Many claim that global warming is obvious and that all arguments against global warming fail. They think that the present temperature increase is part of the larger natural cycle.

The universe has been cooling since the Big Bang. For this reason many strongly believe that the next two centuries are equally as likely to see a cooling trend as they are to see a warming trend. Scientists have not yet reached a consensus on how fast and intense global warming will be. Studies have shown that CO_2 accumulation is the biggest contributor to global warming. However, excess CO_2 can be absorbed by plants and serve as a fertilizer under some conditions.

4.3.5 Kyoto Protocol

Global climate change is an issue where energy questions will be the subject of considerable international political activity. It is an issue that raises key questions about politically sensitive topics such as national sovereignty and international equity. But the Kyoto conference attended by 160 countries in December 1997 and world summits such as the Rio and Montreal meetings showed that most governments feel the need to address the question. According to the Kyoto Protocol, so-called Annex I countries (OECD members plus economies in transition) must reduce their emissions of six greenhouse gases by at least 5% compared with 1990 levels from 2008 to 2012.

According to the Kyoto Protocol:

- Developing countries do not take on commitments to reduce greenhouse gas emissions
- Some developed countries take on commitments to reduce greenhouse gas emissions by 5.2% below 1990 levels by 2008–2012
- It establishes flexibility mechanisms: trading, joint implementation, clean development mechanism.

The Kyoto mechanism foresees:

- Domestic reductions in industrialized countries
- Flexibility mechanisms
- Exchanges among industrialized countries
- Trading: based on greenhouse gas emissions quotas
- Joint implementation: based on projects to mitigate greenhouse gas emissions
- Exchanges between industrialized and developing countries
- Clean development mechanism: based on projects to mitigate greenhouse gas emissions
- Incentives and regulations and public–private partnerships
- Increase of research and development of new technologies
- Energy efficiency and renewable energy
- Greenhouse gas emissions and energy-use taxes.

4.3.6 Carbon Credits

Carbon assessments can play an important role in a strategy to control CO_2 emissions while raising revenue. To limit international emissions of greenhouse gases, carbon assessments have been introduced. Today, fuels and chemicals are predominately derived from unsustainable mineral resources, petroleum and coal, leading to environmental pollution, greenhouse gas emissions, and issues related to energy security. There is consensus among scientists that biomass fuels used in a sustainable manner result in no net increase in atmospheric CO_2. Some would even go as far as to declare that sustainable use of biomass will result in a net decrease in atmospheric CO_2 (Tester *et al.* 1991). This is based on the assumption that all the CO_2 given off by the use of biomass fuels was recently taken in from the atmosphere by photosynthesis. Increased substitution of fossil fuels with biomass-based fuels would therefore help reduce the potential for global warming, caused by increased atmospheric concentrations of CO_2. Biomass can contribute to sustainable development and globally environmental preservation since it is renewable and carbon-neutral. If biomass could be converted into useful energy, the consumption of fossil fuel and greenhouse gas emissions would be decreased.

4.3.7 Carbon Sequestration

Fossil fuels will remain the mainstay of energy production well into the twenty-first century. However, increased concentrations of CO_2 due to carbon emissions are expected unless energy systems reduce the carbon emissions to the atmosphere. To stabilize and ultimately reduce the concentrations of CO_2 gas, it will be necessary to employ carbon sequestration – carbon capture, separation, and storage or reuse. Carbon sequestration, reduced carbon content of fuels, and improved efficiency of energy production and use must play major roles if countries are to enjoy the economic and energy-security benefits which fossil fuels bring to the energy mix.

Specifically, the carbon sequestration process has six elements:

1. Cost-effective CO_2 capture and separation processes
2. CO_2 sequestration in geological formations including oil and gas reservoirs
3. Direct injection of CO_2 into the deep sea and stimulation of phytoplankton growth
4. Improved full life cycle carbon uptake of terrestrial ecosystems
5. Advanced chemical, biological, and decarbonization concepts
6. Models and assessments of cost, risks, and potential of carbon sequestration technologies.

To prevent rapid climate change, it will be necessary to stabilize CO_2 as carbonate by the carbonation process. Carbonation of the widely occurring mineral olivine (Mg_2SiO_4) converts CO_2 into the environmentally benign mineral magnesite ($MgCO_3$) (Bearat *et al.* 2003). In another development, it is proposed that liquid CO_2 is mixed with pulverized limestone and water, and the resulting emulsion is released at a depth below 500 m (Golomb and Angelopoulos 2001).

By accelerating the naturally occurring carbonation of magnesium silicate minerals, one can sequester CO_2 in the geologically stable mineral magnesite. The slow natural geological process that converts both of olivine and serpentine [$Mg_3Si_2O_5(OH)_4$] minerals to magnesite can be accelerated by increasing the surface area, increasing the activity of CO_2 in the solution. Aqueous mineral carbonation reactions take advantage of the natural alteration of ultramafic rocks called serpentinization.

Over geological time, most olivine is eventually converted into serpentine and magnesite, and thus serpentine is more prevalent than olivine. Once magnesite has formed, CO_2 can be stored indefinitely. Reaction rates can be accelerated by decreasing the particle size, raising the reaction temperature, increasing the pressure, and using a catalyst.

When the waters contain dissolved CO_2, magnesite may form as a secondary alteration mineral:

$$2Mg_2SiO_4 + CO_2(g) + 2H_2O \rightarrow Mg_3Si_2O_5(OH)_4 + MgCO_3 + 16.5\,\text{kcal}. \quad (4.4)$$

By increasing the CO_2 activity, one can form magnesite and no serpentine:

$$Mg_2SiO_4 + 2CO_2(g) \rightarrow 2MgCO_3 + SiO_2 + 10.3\,\text{kcal}. \quad (4.5)$$

It is also possible to form calcite by a similar reaction:

$$CaSiO_3 + CO_2(g) \rightarrow CaCO_3 + SiO_2 + 10.6\,\text{kcal}. \quad (4.6)$$

4.4 Effects of Methane Hydrates on Climate Change

Hydrates may affect climate because when they are warmed or depressurized, they decompose and dissociate into water and methane gas, one of the greenhouse gases that warms the planet. Discharge of large amounts of methane into the atmosphere will cause global warming. It has been well documented that methane levels in the atmosphere were lower during glacial periods than during interglacial periods (Chapellaz *et al.* 1993; Blunier *et al.* 1995; Brook *et al.* 1996). In fact, methane is many times more effective as a greenhouse gas than is carbon dioxide. Therefore, if the flux of methane to the atmosphere from dissociating hydrates is of sufficient quantity, this methane can cause global warming. Specifically, as the earth warms, increasing bottom water temperatures could cause gas hydrate disassociation in many marine shelf locations. This gas hydrate disassociation would cause further warming owing to the greenhouse effect of the gas which is released.

Methane is released from hydrates to the atmosphere if the sea level rises. A rise in sea level will result in relatively warm ocean water covering cold Arctic strata. The resulting breakdown of stable gas hydrates within the sediment releases gas into the atmosphere.

Another thing that is unknown about methane hydrates is their stability. If the hydrates should break down, this would cause a massive release of methane into the atmosphere. By some estimates, there is 3,000 times the amount of methane in hydrate formations than is currently found in the atmosphere. Given the size of these deposits and the ability of methane to absorb infrared radiation, this could increase the greenhouse effect dramatically. Further, the decay of these hydrates on the ocean bottom could cause massive landslides, which would cause problems for any extraction facility built nearby. This lack of knowledge about the stability of the hydrate formations prevents us from building any type of extraction facility at this time.

The dissociation of gas hydrates during deglaciation has been linked to the ending of ice ages during the last the last few million years (MacDonald 1990b; Nisbet 1990; Paull *et al.* 1991; Haq 1998; Raynaud *et al.* 1998). Paull *et al.* (1991) suggested that the occurrence of large oceanic gas hydrate reservoirs is the factor limiting the severity of ice ages. During formation of large polar ice sheets, the sea level falls, reducing the pressure on the ocean margin gas hydrates. The shallower gas hydrate deposits become unstable, and release methane into the atmosphere, which causes warming and the ending of the ice age. This scenario remains questionable, however, because almost certainly the permafrost had hydrate reservoirs that were much larger than now during glacial periods, when permafrost regions were much more widespread than they are now.

The scenario of large pulses of methane into the atmosphere might be checked by analyzing methane concentrations in polar ice cores. The methane peak concentrations in the atmosphere would not remain for long, because methane has an atmospheric residence time of only about 10 years, being oxidized to carbon dioxide. The time resolution of ice core sampling has until now not been sufficient to resolve such short-term peaks (Thorpe *et al.* 1996; Raynaud *et al.* 1998).

In addition, gas hydrate dissociation has been suggested as the cause for oceanic anoxia and massive extinctions of marine biota at the end of the Permian (Matsumato 1995): oxidation of the methane within ocean waters could have used up a large part of the dissolved oxygen in the oceans. A much better documented possible episode of massive hydrate dissociation during a short-term warming event occurred at the end of the Paleocene (Dickens *et al.* 1995, 1997; Matsumoto 1995).

Oil and gas wells drilled through permafrost or offshore to reach conventional oil and gas deposits may encounter gas hydrates, which companies generally try to avoid because of a lack of detailed understanding of the mechanical and thermal properties of gas-hydrate-bearing sediments. However, to mitigate the potential hazard in these instances, the wells are cased to separate and protect them from the gas hydrates in the shallower zones as drilling continues deeper. Unless precautions are taken, continued drilling may heat up the sediments surrounding the well bore, causing gas from the dissociated hydrates to leak and bubble up around the casing. Once oil production begins, hot fluids flowing through the well could also warm hydrate-bearing sediments and cause dissociation. The released gas may pool and build up pressure against the well casing, possibly causing damage. Some observers suggest that exploiting the gas hydrate resources by intentionally heating them or by depressurization poses the same risks as drilling through gas hydrates to reach deeper conventional oil and gas deposits (Milkov and Dzou 2007).

The latest Paleocene thermal maximum (LPTM) represents a geologically brief time interval characterized by rapid and profound global warming and associated environmental change. It is marked by a prominent negative carbon isotope excursion of at least –2.5‰ that occurred within 10,000 years, interpreted to reflect a massive and abrupt input of ^{12}C-enriched carbon to the ocean–atmosphere reservoir, possibly as a result of catastrophic gas hydrate release, on time scales equivalent to present-day rates of anthropogenic carbon input. Thermal dissociation of marine gas hydrate and release of massive quantities of methane to the exogenic carbon cycle appears to be the only plausible explanation for this geochemical perturbation (Crouch *et al.* 2001). Methane, after carbon dioxide, is the second most important greenhouse gas in the atmosphere and has caused dramatic climate shifts in the past several million years (Dickens 2000; Hinrichs *et al.* 2003). The LPTM corresponds to important changes in the global distribution of biota, including mass extinction of marine benthic organisms. The dinoflagellate cyst record indicates that surface-dwelling marine plankton in marginal seas also underwent significant perturbations during the LPTM (Crouch *et al.* 2001).

Methane hydrates hold the danger of natural hazards associated with seafloor stability and release of methane to the ocean and atmosphere, and gas hydrates disturbed during drilling are a safety problem. Continental slope instability caused by hydrate decomposition is suggested as a trigger mechanism for underwater landslides and tsunami generation. If large volumes of methane are stored in marine reservoirs, they may significantly influence the sedimentary environment in which they occur. The formation and subsequent decomposition of gas hydrates within the sediments affect the physical properties of the sediment as well. Changes in pressure and temperature will led to decomposition of solid gas hydrates to gas and water, which may lead to sediment instability and failure (Gupta 2004).

Methane is a relatively potent greenhouse gas with a high global warming potential of 72 (averaged over 20 years) or 25 (averaged over 100 years). Methane in the atmosphere is eventually oxidized, producing carbon dioxide and water. As a result, methane in the atmosphere has a half life of 7 years.

Methane concentrations have doubled from 850 ppb to currently approximately 1,750 ppb over the last 150 years (Cicerone and Oremland 1998). In the same time period, carbon dioxide increased from 278 to 365 ppm. There is a large, but unknown, amount of methane in methane hydrates in the ocean floors. The earth's crust contains huge amounts of methane. Large amounts of methane are produced anaerobically by methanogenesis. Other sources include mud volcanoes, which are connected with deep geological faults, and livestock, from enteric fermentation. Methane has 21 times the global warming potential as the same mass of carbon dioxide (St. Louis *et al.* 2000). Anthropogenic inputs, such as rice paddies, livestock, and biomass combustion, contribute the largest proportion to the atmospheric concentration (71%) (Reeburgh 1996). Natural sources, such as wetlands and termites, contribute, along with other small sources, approximately 29%. Recently, it has been suggested that up to 18% of the atmospheric methane is emitted from reservoirs (St. Louis *et al.* 2000).

Atmospheric methane concentrations, however, would be even much higher if the huge methane pools that are stored in ocean and lake sediments were released to the atmosphere. Current research has therefore focused on the contribution of methane from gas hydrates, a phenomenon that was overlooked some 20–30 years ago (Kvenvolden 1988a). Higher temperatures from global warming lead to the dissociation of methane hydrates and the subsequent release of methane to the water column. It is estimated that twice the amount of energy is stored in gas hydrates than in all combined coal, gas, and oil deposits (Kvenvolden 1998).

It is assumed that the release of methane from marine hydrates during climatic maxima and minima has played a significant role in climate change. The earth has witnessed several intervals of climate change, typified by lowering of and rises in the sea level due to rapid cooling and warming. Because several of these time intervals are characterized by major inputs of carbon to the ocean and atmosphere, and because the change in pressure–temperature conditions during these time slices may have reduced the stability of gas hydrate reservoirs, it is believed that the release of methane from marine hydrates has played a significant role in warming the earth's climate (Milkov and Etiope 2005). During glaciation, when the sea level drops, large volumes of methane may be released; conversely, when the sea level rises, the lower limit of gas hydrate stability migrates downward and more gas may be trapped. Sea-level-controlled methane releases could act as a negative feedback to advancing glaciation by producing global warming and as a positive feedback during sea-level rise (Gupta 2004).

During the formation of gas hydrates, methane and water become immobilized within the sediment pore spaces. Because of the presence of these solids, the sediment cannot become consolidated because the water cannot be expulsed with increasing overburden as more sedimentation occurs. Cementation of the sediments does not occur when pore spaces are filled with hydrates rather than with water, from which minerals such as calcite can be precipitated. Gas-hydrate-rich sediments are thus cemented by the hydrates, which may occupy much of the sedimentary section, but which are not stable when the temperature rises or the pressure falls. This may lead to problems during continued sedimentation and further burial of the gas hydrates: the hydrates will become buried so deeply that the temperature will increase according to the regional geothermal gradient. The hydrates will then no longer be stable, and will disintegrate into a liquid water and gas mixture. The basal zone of the gas hydrate becomes underconsolidated, possibly overpressured because of the release of the methane, leading to the development of a zone with low shear strength where failure could be triggered and massive landslides could occur. With the landslides, more gas could escape (Bakker 1998).

Several examples of extremely large slumps possibly linked to gas hydrates have been described, e.g., on the Norwegian continental margin (Bugge *et al.* 1987), where debris from the giant, three-part Storegga slide, over 450 m thick, is spread over a distance of 800 km. One of the Storegga slides caused a tsunami to deposit sediment up to 4 m above the high water line in Scotland (Nisbet and Piper 1998). There are more of these megaslides in the same region (Laberg *et al.* 2000).

Gas hydrates are a significant hazard for drilling and production operations. Gas hydrate production is hazardous in itself, as well as for conventional oil and gas activities that place wells and pipelines into permafrost or marine sediments. For activities in permafrost, two general categories of problems have been identified: (1) uncontrolled gas releases during drilling and (2) damage to the well casing during and after installation of a well. Similar problems could occur during offshore drilling into gas-hydrate-bearing marine sediments. Offshore drilling operations that disturb gas-hydrate-bearing sediments could fracture or disrupt the bottom sediments and compromise the well bore, pipelines, rig supports, and other equipment involved in oil and gas production from the seafloor (Folger 2008). The problems may differ somewhat between onshore and offshore operations, but they stem from the same characteristic of gas hydrates: decreases in pressure and/or increases in temperature can cause the gas hydrate to dissociate and rapidly release large amounts of gas into the well bore during a drilling operation (Kvenvolden 1991; Sloan 1998b).

The earth's atmosphere has a large number of sources and sinks for methane, including gas hydrates, which exist in metastable equilibrium with the environment. This equilibrium is affected by natural changes in pressure and temperature. The amount of methane that is trapped in gas hydrates is perhaps 3,000 times the amount in the atmosphere and this enormous quantity of methane is available for release, with catastrophic consequences for the global climate (Desa 2001).

The rapid climate change in the past from the glacial to the interglacial period is indeed attributed to the huge release of methane from gas hydrates. Some of the models (Kenett *et al.* 2000; Blunier 2000) linked these observations with climatic change caused by hydrate decomposition or *vice versa*. Accordingly, glaciation and/or sea-level fall will reduce the hydrostatic pressure on the sediments of the shelf and slope at mid-latitudes, thereby destabilizing hydrates and consequently releasing methane into the atmosphere, The methane so released will cause rapid warming, which will lead to dissociation of hydrates in the permafrost regions and shallow continental margins of the high-latitude regions, a positive feedback causing further warming. Consequently, ice in the higher latitude will melt, thereby increasing the sea level. This rise in sea level will in turn increase the hydrostatic pressure on the sediments of the shelf and slope at mid-latitudes, initiating a negative feedback that inhibits further release of methane from hydrates.

Submarine sediment slumping or sliding occurs when huge piles of unstable sediments build up. In some cases, movement of these sediment slumps or slides can possibly be caused by the decomposition of gas hydrates and the resulting expansion or release of gas. In the continental slope area, continued sedimentation may lead to burial of hydrated sediment to such a depth that the hydrates are no longer stable and dissociate into a liquid–gas–water mixture. The dissociated hydrates create a zone of weakness in the sediment where slope failure could be triggered by further gravitational loading or earthquakes. These changes, though of long duration, may have caused solid sediment to become gas-cut mud and resulted in mud diapers, mud volcanoes, and mud slide flows, depending on the sediment composition and bottom topography (Desa 2001).

Methane hydrates are located in the shallow submarine geosphere, which is a finely balanced system in equilibrium with all its components, such as sediment, pore water, fluid flows, pressure, temperature, overlying water, and hydrate. Removal of any one component of this equilibrium may destabilize the whole system, leading to irreparable damage. The destabilizing factors may be either natural perturbations or perturbations associated with exploitation. Studies have indicated that methane hydrates have the potential to affect global climate and the geological environment on a catastrophic scale.

4.5 World's Methane Hydrate Resources

There are several challenges to commercially exploiting gas hydrates. How much and where gas hydrate occurs in commercially viable concentrations are not well known, and how the resource can be extracted safely and economically is a current research focus. Estimates of global gas hydrate resources, which range from at least 100,000 TCF to possibly much more, may greatly overestimate how much gas can be extracted economically. Reports of vast gas hydrate resources can be misleading unless those estimates are qualified by the use of terms such as "in-place resources," "technically recoverable resources," and "proved reserves" (Folger 2008):

1. The term "in place" is used to describe an estimate of gas hydrate resources without regard for technical or economical recoverability. Generally, these are the largest estimates.
2. Undiscovered technically recoverable resources are producible using current technology, but this does not take into account economic viability.
3. Proved reserves are estimated quantities that can be recovered under existing economic and operating conditions.

For example, the US Department of Energy's Energy Information Agency estimates that total undiscovered technically recoverable conventional natural gas resources in the USA are approximately 1,300 TCF, but proved reserves are only 200 TCF. This is an important distinction because there are no proved reserves for gas hydrates at this time. Gas hydrates have no confirmed past or current commercial production.

The earlier Kvenvolden (1988b) estimate represents extrapolation of an estimate of the hydrate present off northern Alaska to all continental margins. The latest Kvenvolden and Rogers (2005) estimate takes into account the most recent work in the field, providing a constrained range and a "consensus" central estimate. The theoretical maximum volumes have been estimated by assuming that methane hydrate occurs at all locations where conditions are favorable and that it is fully saturated; the result, 2,700,000 TCF in oceanic deposits and 12,000,000 TCF in continental deposits, is unlikely to represent the actual case (Trofimuk *et al.* 1977; Kvenvolden 1981; McIver 1981; Meyer 1981; MacDonald 1990b; Gornitz and Fung 1994).

The Department of the Interior's US Geological Survey and Minerals Management Service has reported only in-place estimates of US gas hydrate resources. However, a November 12, 2008 US Geological Survey estimate of undiscovered technically recoverable gas hydrates in northern Alaska probably represents the most robust effort to identify gas hydrates that may be commercially viable sources of energy. Despite the lack of a production history, the US Geological Survey report cites a growing body of evidence indicating that some gas hydrate resources, such as those in northern Alaska, might be exploited with existing technology despite only limited field testing (Folger 2008).

The extent of worldwide gas hydrate occurrences has been evaluated using seismic exploration, because gas hydrates are characterized by the occurrence of an anomalous reflector parallel to the seafloor (bottom-simulating reflector, BSR). But BSRs have failed to locate gas hydrate horizons in a few cases, e.g., at Ocean Drilling Program site 994C located on the Blake Ridge, North Atlantic, where much information comes from the geochemical and sediment parameters (Hornbach *et al.* 2003). Besides, microfaunal groups such as benthic foraminifera can also provide useful information for locating the gas hydrate horizons. Some species of benthic foraminifera colonize hydrocarbon-seeped bacterial mats and may be attracted to methane gas or hydrogen sulfide gas emissions.

Methane hydrates are widespread in sea sediments hundreds of meters below the seafloor along the outer continental margins and are also found in Arctic permafrost. Some deposits are close to the ocean floor and at water depths as shallow as 150 m, although at low latitudes they are generally only found below 500 m. The deposits can be 300–600 m thick and cover large horizontal areas. A nearby deposit nearly 500 km in length is found along the Blake Ridge off the coast of North Carolina at depths of 2,000–4,000 m.

Gas hydrates trapped in the marine sediments require multifaceted, extensive efforts to bring them into world energy balance. Government organizations, laboratories, and the oil industry have recently undertaken projects to explore gas hydrates in Indian offshore areas. However, participation by academic institutions is negligible owing to lack of collaboration among them. There should be greater involvement of researchers from academic institutions to intensify efforts in the exploration of gas hydrate horizons. Besides, the USA needs more sophisticated drilling vessels with deep-penetration drilling devices and a repository to store sediment cores (Gupta 2004). Efforts to explore the development of gas hydrate resources are already under way in energy-resource-poor countries such as Japan and India.

The US Department of Energy's Gulf of Mexico Joint Industry Project was an aggressive multimillion dollar gas hydrate research program focused on the US Gulf of Mexico. Joint Industry Project participants included the US Department of Energy and a group of petroleum industry companies, including ConocoPhillips, Halliburton, and Japan National Oil Company. This multiyear program was summarized by Shirley (2004).

Another program of note in 2004 was that undertaken by a Japanese-government-sponsored gas hydrate research organization: the Research Consortium for Methane Hydrate Resources in Japan (also known as the MH21 Research Consortium). The MH21 Research Consortium gas hydrate research extends beyond coastal Asia. In 2002, production testing of gas hydrates in the Mackenzie Delta (Canada) was conducted by an international consortium that included Japan National Oil Company and the Geological Survey of Canada. Detailed results of that project were publicly presented at a conference in Chiba, Japan, in December 2003.

Most methane hydrate deposits in the USA are located in the Alaskan outer continental shelf. Additional deposits are onshore in northern Alaska, in the Gulf of Mexico, and on the western and eastern outer continental shelves. The US Geological Survey estimates the USA to have about 200,000 TCF of methane hydrate. This number dwarfs the estimated 1,400 TCF of conventional methane recoverable from natural gas reserves and reservoirs. Worldwide estimates of methane hydrate deposits reach the overwhelming number of 400,000,000 TCF – far outdistancing the 5,500 TCF of proved worldwide gas reserves.

The USA relies on natural gas for heating, cooking, transportation, industry, and manufacturing. As national and worldwide supplies become scarcer, new resources for methane are a domestic priority. According to the US Department of Energy, Office of Fossil Energy, "If only one percent of the methane hydrate resource could be made technically and economically recoverable, the USA could more than double its domestic natural gas resource base." To that end, Congress approved the Methane Hydrate Research and Development Program in 2000. This program has provided tens of millions of dollars in matching seed funds for research and development of methane hydrate as an energy resource. More recently, in November 2005, the US Department of Energy announced \$ 2 million in funding for five research projects focused on the energy potential, safety, and environmental impacts of methane hydrate exploration and development. As natural gas supplies decrease and the demand for cleaner fuels increases, methane hydrates may well play a role in the US energy portfolio.

The US Geological Survey estimates that there is twice the amount of carbon to be found in methane hydrate deposits as there is in all other fossil fuels combined. However, this estimate was made with scant information, and could very well be wild speculation. There are better data for the existence of methane hydrates in certain locations. Mappings by the US Geological Survey of offshore North Carolina and South Carolina have revealed the possible existence of a 1,300-TCF methane hydrate deposit. If this is true, and if it could be extracted safely, it would represent a 700% increase in the current natural gas deposits in the USA. At current consumption rates, this would be a 70-year reserve of natural gas. Figure 4.5 shows the locations of known and expected concentrated methane hydrate deposits.

Globally, the amount of gas hydrate to be found offshore along continental margins probably exceeds the amount found onshore in permafrost regions by 2 orders of magnitude, according to one estimate. With the exception of the assessments discussed above, none of the global gas hydrate estimates are well defined, and all are speculative to some extent (Folger 2008). One way to depict the potential size and producibility of global gas hydrate resources is by using a re-

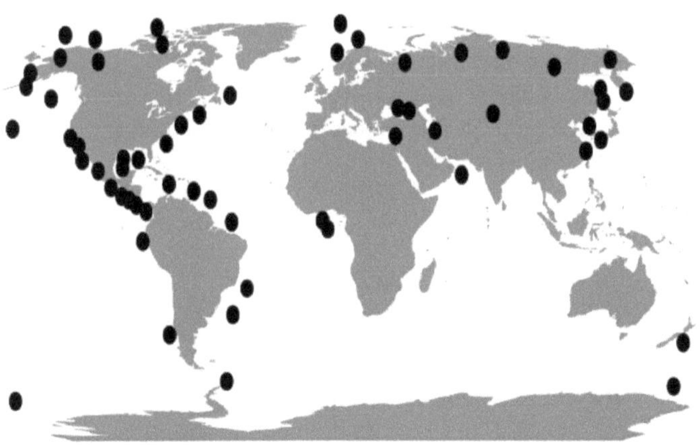

Fig. 4.5 Locations of known and expected concentrated methane hydrate deposits (Kvenvolden 1988c)

source pyramid (Boswell and Collett 2006). The apex of the pyramid shows the smallest but most promising gas hydrate reservoir, which may host tens to hundreds of TCF. The bottom of the pyramid shows the largest but most technically challenging reservoir.

Sandstones are considered superior reservoirs because they have much higher permeability than shales, which can be nearly impermeable. The marine shale gas hydrate reservoir may host hundreds of thousands of TCF, but most or all of that resource may never be economically recoverable. It is likely that continued research and development efforts in the USA and other countries will focus on producing gas hydrates from Arctic and marine sandstone reservoirs.

4.5.1 Gas Hydrates on the North Slope, Alaska

On November 12, 2008, the Washington Post reported that the US Geological Survey would issue a resource assessment of Alaska's gas hydrates "estimating that 85.4 trillion cubic feet of natural gas can be extracted … an amount that could heat more than 100 million average homes for more than a decade." This does not mean that commercial production of these resources is imminent, but it does suggest there is a vast natural gas resource trapped in Alaska's gas hydrates (Collett 1993).

The US Geological Survey assessment indicates that the North Slope of Alaska may host about 85 TCF of undiscovered technically recoverable gas hydrate resources. According to the report, the technically recoverable gas hydrate resources could range from a low of 25 TCF to as much as 158 TCF on the North Slope. Total US consumption of natural gas in 2007 was slightly more than 23 TCF. Of the mean estimate of 85 TCF of technically recoverable gas hydrates on the North Slope, 56% are located on federally managed lands, 39% on lands and offshore waters managed by the State of Alaska, and the remainder on Native lands (Collett *et al.* 1989; Collett 1997). The total area covered by the US Geological Survey assessment is 55,894 square miles, and extends from the National Petroleum Reserve in the west to the Arctic National Wildlife Refuge in the east. The area extends north from the Brooks Range to the state–federal offshore boundary 3 miles north of the Alaskan coastline (Folger 2008).

4.5.2 Gas Hydrates in the Gulf of Mexico

On February 1, 2008, the Minerals Management Service released an assessment of gas hydrate resources for the Gulf of Mexico. The report gives a statistical probability of the volume of undiscovered in-place gas hydrate resources, with a mean estimate of over 21,000 TCF. The Minerals Management Service report estimates how much gas hydrate may occur in sandstone and shale reservoirs, using a combination of data and modeling, but does not indicate how much is recoverable with current technology (Milkov and Sassen 2001). The report notes that porous and permeable sandstone reservoirs have the greatest potential for actually producing gas from hydrates, and gives a mean estimate of over 6,700 TCF of sandstone-hosted gas hydrates, about 30% of the total mean estimate for the Gulf of Mexico (Milkov and Sassen 2003). Even for sandstone reservoirs, however, the in-place estimates for gas hydrates in the Gulf of Mexico likely far exceed what may be commercially recoverable with current technology (Sassen *et al.* 2001). The Minerals Management Service is planning similar in-place gas hydrate assessments for other portions of the US outer continental shelf, including Alaska (US-OCSR 2008).

Models of gas hydrate stability for the northern Gulf of Mexico continental slope address basic problems of gas hydrate geology. Resource estimation is based on assessment of the volume of the gas hydrate stability zone and concentration in sediments. The total volume of gas trapped in gas hydrate in the Gulf of Mexico is estimated to be 100 times less than previously predicted. However, structural-control-led accumulations of gas hydrate on the rims of salt withdrawal basins could be economic in the future. Bacterial gas hydrates in salt withdrawal basins are unlikely to represent a significant energy resource because they are disseminated (Milkov and Sassen 2003).

4.5.3 Gas Hydrates in Russia's West Siberian Basin

The Messoyakha Field is located in permafrost terrain on the eastern margin of Russia's West Siberian Basin; it was developed as a conventional gas field and produced continuously from 1970 through 1978 and thereafter intermittently, primarily in the summer to accommodate regional industrial demand. As is normally the case, reservoir pressure declined as a consequence of production; however, the reservoir pressure remained substantially higher than normally expected. A 100-m-thick methane hydrate zone is located 700 m beneath the surface, and the apparent difference between the actual and predicted pressure decline behavior has been attributed to recharging of the reservoir with gas derived from pressure-decline-induced decomposition of the natural gas hydrates in this overlying layer. In 1990, the gas evolved from it reportedly comprised nearly half of the cumulative field production, although some investigators have expressed doubt that gas hydrate production actually occurred.

4.5.4 Gas Hydrates in Canada

Scientific field studies of natural gas hydrates have been carried out in Canada for many years by the Geological Survey of Canada and by a number of universities.

The two largest programs are (1) Arctic permafrost-related hydrate studies, including the Mallik research drilling on the Mackenzie Delta, and (2) geophysical studies and Ocean Drilling Program drilling of marine gas hydrate beneath the continental slope off Canada's west coast.

Assessments of gas hydrate occurrence in the Mackenzie Delta–Beaufort Sea region have been made mainly on the basis of data obtained from hydrocarbon exploration wells. A database produced by Smith and Judge (1995) showed 17% of 145 onshore wells in the Mackenzie Delta area having evidence of gas hydrate. Smith and Judge (1995) estimated that the Mackenzie Delta–Beaufort Sea region may contain 16 trillion cubic meters of natural gas in hydrate form.

Marine gas hydrate was first discovered beneath the continental slope off the west coast in conventional marine multichannel seismic data acquired as part of the Geological Survey of Canada's Frontier Geoscience Program. This program involved surveys of the resources in the 200-mile zone off Canada's west coast following the Law of the Sea convention. Several additional multichannel surveys were acquired as part of a site survey for drilling by the international Ocean Drilling Program. By region, the estimated in-place resource ranges are:

- McKenzie Delta–Beaufort Sea: 0.24×10^{13}–$8.7 \times 10^{13}\,m^3$
- Arctic Archipelago: 0.19×10^{14}–$6.2 \times 10^{14}\,m^3$
- Atlantic Margin: 1.9×10^{13}–$7.8 \times 10^{13}\,m^3$
- Pacific Margin: 0.32×10^{14}–$2.4 \times 10^{14}\,m^3$.

In December 2003, the Canadian Mallik 2002 Gas Hydrate Production Research Well Program partners (including the US Geological Survey and the US Department of Energy) publicly released the results of the first modern, fully integrated field study and constrained production test of a natural gas hydrate accumulation. The Mallik 2002 gas hydrate production testing and modeling effort has, for the first time, enabled rational assessment of the production response of a gas hydrate accumulation. Project-supported gas hydrate production simulations have shown that under certain geological conditions gas can be produced from gas hydrates at rates exceeding several million cubic feet of gas per day (Collett 2004).

4.5.5 Projects Related to Methane Hydrates in Japan, India, China, Malaysia, and Korea

Methane hydrates are found in the Nankai Trough located offshore Japan. The Nankai Trough runs along the Japanese islands, where forearc basins and accretionary prisms developed extensively and BSRs have been recognized widely. High-resolution seismic surveys in 1997, 2001, and 2002 and drilling the Nankai Trough wells conducted by the Ministry of Economy, Trade and Industry have revealed the subsurface gas hydrate widely distributed at a depth interval from 200 to 270 m below the seafloor.

There were major drilling campaigns in the Nankai Trough offshore Japan in 2004 (Fujii *et al.* 2005), and in several widely spaced sites offshore India in 2006. Unlike the US-led efforts, these programs had the specific objective of assessing the fossil fuel potential of gas hydrate. Comprehensive reports of these campaigns are not yet in the public domain, so there is little public record available to assess the efficacy of exploration paradigms. However, these programs serve in some respects as models of national resource assessment. The national project of methane hydrate development in Japan (MH21 Research Consortium) started in 2001. In the project operated by the MH21 Research Consortium, a lot of experimental and computational research and development was carried out on modeling and the production method for methane hydrates (MH21 Research Consortium 2004).

China has recently announced a $ 100 million program of hydrate exploration. The state-owned oil companies of Malaysia and Korea have signaled their intention to explore for hydrate in their territorial waters.

The main technology barrier is the lack of validated means of reliably finding significant marine gas hydrate resources. A multisite geological and geophysical exploration program, followed up by a multisite drilling campaign, is needed to accelerate the assessment of marine gas hydrate as an energy resource.

4.5.6 A Case Study: Methane Hydrates in the Black Sea

The potential of gas hydrates as a source of methane is being investigated in the Black Sea. The Black Sea is the world's most isolated sea. It is a unique energy-rich sea. It contains gas hydrates and hydrogen sulfide in abundance as methane and hydrogen sources, respectively. The Black Sea is the world largest water body containing hydrogen sulfide; the hydrogen sulfide layer begins about 200 m below the surface. The concentrations of hydrogen sulfide range from 1.5 to 2.3 mL/L in the Black Sea at depths of 200–1,500 m (Eremeev *et al.* 1998). The Black Sea has suitable conditions (pressure and temperature) for the formation of natural gas hydrates. On the other hand, the anaerobic but hydrogen sulfide conditions of the Black Sea, starting at a depth of 200 m from sea level, make it more suitable for hydrate formation.

Methane hydrates can form in the Black Sea at water depths exceeding 580–700 m. One of the first reported observations of gas hydrates in marine sediments was from the Black Sea (Yefremova and Zhizhchenko 1974). The area of the Black Sea that is suitable for gas hydrate formation is evaluated at 288,100 km^2, representing about 68% of the total Black Sea or almost 91% of the deepwater basin. Methane has discharged into the Black Sea and the global ocean via fluid flow through submarine mud volcanoes. The areas of active fluid venting and mud volcanism were investigated in the Black Sea below the oxic zone at depths between 800 and 2,200 m.

The Black Sea has a surface area of 423,000 km^2, a volume of 5.45×10^5 km^3, and a maximum depth of 2,212 m, making it the largest anoxic water basin in the world. Freshwater river inflow resulted in lower surface water salinity (17.5–18.5‰), whereas the deepwater salinity is 22.3‰ (Murray 1991).

Currently, the Black Sea is the largest anoxic water body on the planet and represents the closest contemporary analog to past sulfidic oceans (Overmann and Manske 2006). Figure 4.6 shows the average methane concentrations in the water column in the Black Sea. The Black Sea has a layered structure where only the upper 150 m is oxic, whereas the main water body is anoxic. Methane seepage is extremely intense on the shelf and on the slope of the Black Sea. The methane content increases from the water surface with depth: in the central Black Sea, the concentration has been measured as 0.036 μM in the 150-m surface layer, increasing from 5.1 to 13.9 μM at 250 and 550 m, respectively, and reaching the plateau value at the bottom (Overmann and Manske 2005; Popescu *et al.* 2006; Blumenberg *et al.* 2007).

A mud volcano area in the deep waters (below 2,000 m) of the Black Sea was studied by hydroacoustic measurements during several cruises between 2002 and 2004 (Greinert *et al.* 2006). The recorded echograms show that bubbles rise about 1,300 m high through the water column, to a final water depth of about 770 m, which is approximately 75 m below the phase boundary of pure methane hydrate in the Black Sea. To better understand how a methane bubble (gas analyses of the pore water and gas hydrate gave 99.4% methane) can rise so high without dissolving, a recently developed bubble dissolution model can be applied that takes into account a decreased mass transfer due to an immediately formed gas-hydrate rim.

A peculiar type of methane-driven microbial habitat is the large gas seepage area in the anoxic part of the northwestern slope of the Crimean Peninsula in the Black Sea (Michaelis *et al.* 2002). The deep anoxic shelf of the northwestern Black Sea has numerous gas seeps, which are populated by methanotrophic microbial mats in and above the seafloor. Above the seafloor, the mats can form tall

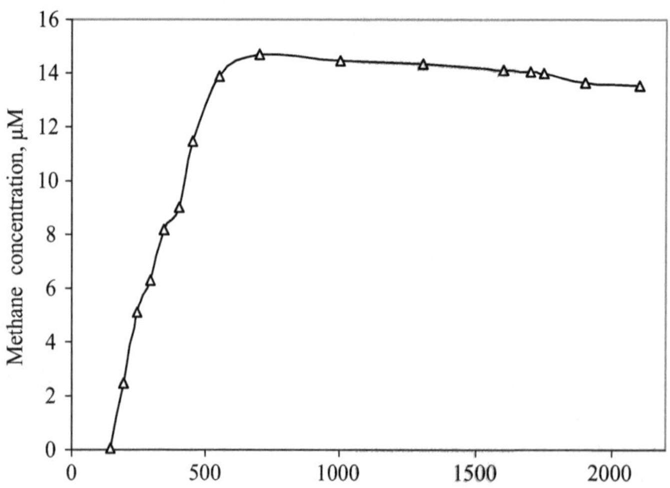

Fig. 4.6 Average methane concentrations in the water column in the Black Sea

reef-like structures composed of porous carbonate and microbial biomass (Mahajan *et al.* 2007). Reduction of carbon dioxide to methane was investigated in the presence or absence of methane, sulfate, and hydrogen. The level of carbon dioxide reduction was about 10% of the level of anaerobic oxidation of methane. However, since considerable methane formation was observed only in the presence of methane and sulfate, the process appeared to be a re-reaction of anaerobic oxidation of methane rather than net methanogenesis (Treude *et al.* 2007). Massive microbial mats covering up to 4-m-high carbonate buildups prosper at methane seeps in anoxic waters of the northwestern Black Sea shelf. Strong ^{13}C depletions indicate the incorporation of methane carbon into carbonates, bulk biomass, and specific lipids. Obviously, anaerobic microbial consortia could have generated both carbonate precipitation and substantial biomass accumulation during earlier periods of Earth's history (Michaelis *et al.* 2002).

Large amounts of dissolved methane (more than 98%) are effectively oxidized by anaerobic and aerobic microorganisms in the Black Sea water column. In the oxic–anoxic transition zone, methane concentrations rapidly decrease and $^{13}CH_4$ enrichment is observed. The data indicate that aerobic rather than anaerobic methanotrophs are responsible for the strong $^{13}CH_4$ enrichment occurring in suboxic waters within the oxic–anoxic transition zone of the Black Sea water column (Blumenberg *et al.* 2007).

During a transition period from oxic to anoxic conditions in the bottom water, rates of sulfate reduction and methane production, methane fluxes, as well as concentration profiles of sulfate, sulfide, and methane were measured in sediments at the seabed. During the time period of the investigation, methane and sulfide concentrations increased, whereas sulfate and oxygen concentrations declined exponentially in the deep water (Piker *et al.* 1998; Weber *et al.* 2001).

4.6 Properties of Methane Hydrates

Methane hydrates are common in sediments deposited on high-latitude continental shelves and at the slope and rise of continental margins with high bioproductivity (Kvenvolden 1988b). High biological production provides the organic matter buried in the sediment, which, during early diagenesis and after exhausting oxygen, sulfate, and other electron acceptors, eventually generates methane through fermentative decomposition and/or microbial carbonate reduction (Suess 2002). The properties of sediment-hosted gas hydrates are strongly determined by the texture, structure, and permeability of the sediment and the mode of supply of methane.

According to the well-known hydrate model, the water molecules from a well-defined crystal lattice (the host lattice) containing cavities into which small gas molecules (guests) may be adsorbed; under appropriate conditions the adsorption energy may then reduce the free energy of the hydrate sufficiently to make the hydrate phase more stable than either pure water or ice (Rodger 1990). In the hydrate crystal lattice, guest–guest interactions are negligible. The host lattice is considered merely to give rise to an environment in which the guest molecules evolve, and the thermodynamic properties of the hydrate result from the classical behavior of individual guest molecules within the cavity potential. The hydrate formation process was described by gas absorption, primary and secondary nucleation, growth, agglomeration, and breakage (Kvenvolden 1995).

The interaction between hydrate and the host sediment at the grain level has been highlighted in recent years as laboratory and analytical investigations have shown that hydrate is not restricted to forming in a unique way in the pore space (Clennell *et al.* 1999; Tohidi *et al.* 2001; Waite *et al.* 2004). As gas hydrates become an ever more important area of interest, there is need for greater knowledge of the effects that gas hydrate morphology can have on the host sediment properties to correctly determine the amounts in the subsea (Kingston *et al.* 2008). Methane hydrate has a porous nature. Diffusion of gas and water through hydrate has been observed at the grain level (Tohidi *et al.* 2001). When formed by rapidly cooling sediment, hydrate preferentially grows at the gas–water interface, and porous hydrate is formed.

The macroscopically dense hydrates contain nanopores, whereas the macroscopically porous hydrates have no nanopores or only remnants. The variability in pore configuration and volume leads to considerably lower bulk densities of methane hydrates than expected from theoretical and experimental considerations. The presence of hydrate, which replaces water in the pore space, strongly alters the physical properties of the sediments in which it occurs. The bulk densities of methane hydrates range from 0.35 to 0.75 kg/L and are inversely correlated with the pore volume, which ranges from 10 to 70% by volume. The density of methane hydrate is lower than that of seawater (Suess *et al.* 2002). The average hydrate composition is 1 mol of methane for every 5.75 mol of water. In general, the observed density is around 0.9 kg/L. One liter of methane clathrate solid would contain 168 L of methane gas at standard temperature and pressure.

Experiments have shown the anomalous preservation of methane hydrate at temperatures below 273 K at ambient pressure with simultaneous formation of an ice phase at temperatures above 242 K (beginning of methane hydrate dissociation). When the pressure on a sample of methane hydrate was rapidly decreased under the equilibrium curve, the preservation of up to 93% of methane was observed (Stern *et al.* 2001; Subbotin *et al.* 2006). The experimental pressure of methane hydrate decomposition is about 13 bar at 250 K. The unusual behavior of some gas hydrate on heating, i.e., delayed dissociation of gas hydrate at temperatures above equilibrium for hydrate decomposition but below the water melting point, was the subject of many experimental investigations in the last few years (Hesselbo et al. 2000). Such interest is caused by possible application of this effect for production, storage, and transportation of natural gas. It was shown that the model of the ice shield cannot explain anomalous preservation of big amounts of methane hydrate. A most popular explanation of the self-preservation is kinetic handicapping of methane diffusion through an ice film, but the exact mechanism of the self-preservation effect in gas hydrate is not clear yet. It was found that the density in the hydrate phase is greater than in the ice phase, which is in agreement with experimental data. The anomalous behavior of the local density at the interface is apparently connected to the increased pressure there (Subbotin *et al.* 2006).

Parameters affecting gas hydrate formation and dissociation include temperature, pore pressure, gas chemistry, and pore water salinity. Any change in the equilibrium of these parameters may result in dissociation and/or dissolution of the gas hydrate. A series of empirical expressions for predicting gas hydrate stability, its volume fraction out of the pore space, and its mass density were established in different systems with consideration of gas composition (CH_4, C_2H_6, H_2S), salinity (NaCl, seawater), and pore size at temperatures between 273.15 and 300 K, based on a gas hydrate thermodynamic model (Sultan *et al.* 2004). At higher temperatures, the empirical predictions for gas hydrate stability do not reproduce the experimental data well, because they are based on a van der Waals model. The empirical expressions for predicting gas hydrate stability in the systems $CH_4 + H_2S + H_2O$, CH_4 + seawater + pore size, $CH_4 + H_2S$ + NaCl, and $CH_4 + CO_2$ + NaCl, and for evaluating the gas hydrate fraction and its density need further validation owing to lack of available published experimental data. However, the empirical expressions for the gas hydrate fraction and its density show that the effects of pore size and salinity are negligible; the gas hydrate fraction will increase if the methane concentration continuously increases relatively in excess of the methane solubility and decreases with pressure within the gas hydrate stability zone (Sloan 1998; Sultan 2007).

Natural gas hydrates may be of biogenic origin (as a result of decay of organic matter and methane formation – methanogenesis) or thermogenic origin. A clue to the origin may by found in their structure. Biogenic gas hydrates are of a type known as sI, whereas thermogenic hydrates are of sII and sH type.

4.7 Methane Hydrate Stability

Under suitable conditions of low temperature and high pressure, a gas molecule will react with water to form hydrates according to

$$G + N_H H_2O = G N_H\, H_2O,$$

where G is a gas molecule and N_H is the hydration number. Of particular interest are hydrates formed by hydrocarbon gases when the gas molecule is an alkane, especially methane, in which case $N_H = 6$, and 1 vol of hydrates contains about 164 vol (at standard temperature and pressure) of gas Natural hydrates in geological systems also include CO_2, H_2S, and N_2 as guests. Natural hydrate deposits involve mainly methane, and occur in two distinctly different geological settings where the necessary low temperatures and high pressures exist for their formation and stability: in the permafrost and in deep ocean sediments.

Gas hydrates are stable under low temperatures and high pressures. The hydrate stability zone in marine environments is a function of the water depth, the seafloor temperature, and the geothermal gradient. Any changes to the temperature and/or pressure, both at the surface and in the area adjacent to the hydrate, will affect the thickness of the stability zone. Although temperature and pressure are the main influencing factors in the formation of gas hydrates and the thickness of the hydrate stability zone, other factors, such as gas chemistry and gas availability, will also alter the thickness and location of the hydrate stability zone (Lerche and Bagirov 1998; Nixon and Grozic 2006).

The growth and stability of the free-gas zone beneath gas-hydrate-related BSRs has been investigated using analytical and numerical analyses to understand the factors controlling the formation and depletion of free gas. Gas forms across a thick zone because the upward fluid flux is relatively low and because the gas–water solubility decreases to a minimum several hundred meters below the seabed (Haacke *et al.* 2008). Methane sI hydrates are stable at 1 atm and 268.2 K and methane sII hydrates are extraordinarily stable at 1 atm and 268.2–270.2 K. Test innovations that achieved ultrastabilities give insight into hydrate self-preservation mechanisms. The resulting hydrate stability window is 268.2–270.2 K at 1 atm. Methane sI, as well as natural gas sII, hydrates exhibit only minimal decomposition upon reduction of the confining system pressure to 1 atm in the 268.2–270.2 K stability window. The total gas that evolved after 24 h at 1 atm in the stability window typically amounted to less than 0.5% of the originally stored gas, and this ultrastability was shown to persist when the test was allowed to run for 256 h before being terminated. The entire methane sI or natural gas sII hydrate mass remains stable during pressure reduction to 1 atm, whereas previous reports defined hydrate anomalous stability for only about 50% of the fractional hydrate remnants (Zhang and Rogers 2008).

Figure 4.7 shows a phase equilibrium diagram for methane hydrates. At pressures and temperatures outside the hydrate stability range, melting and decomposition of gas hydrates will occur. Decomposition will result in the release of water and methane gas, but requires input of heat. As decomposition occurs, the gas and water released cause a volume expansion. If the heat transport and the pressure change processes are fast compared with the pore pressure dissipation processes, the excess pore pressure and the reduction in the effective stress can be estimated

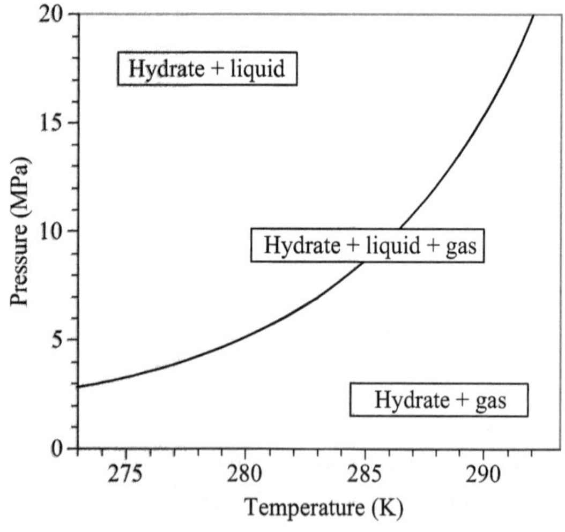

Fig. 4.7 Phase equilibrium diagram for methane hydrates

(Nixon and Grozic 2006). The decomposition of gas hydrates can stem from any change in the pressure and temperature regime in the hydrate stability zone and results in a significant volume change.

In a natural environment, the importance of the pore pressure generated by hydrate dissolution and dissociation depends on the rate of temperature increase and on the ratio of thermal diffusivity to hydraulic diffusivity. For a slow rate of temperature increase and a thermal diffusivity 2–3 orders of magnitude higher than the hydraulic diffusivity, the excess pore pressure generated by both hydrate dissolution and hydrate dissociation is expected to be low (Sultan *et al.* 2004). However, hydrate dissolution and dissociation may significantly alter the structure and mechanical properties of the marine sediments, and the subsequent softening and decrease of the shear strength is probably the main driving factor of sediment deformations and slope instabilities.

Figure 4.8 shows a phase boundary diagram demonstrating the gas hydrate/freshwater stability field. Boundaries are given for the pure methane/pure water

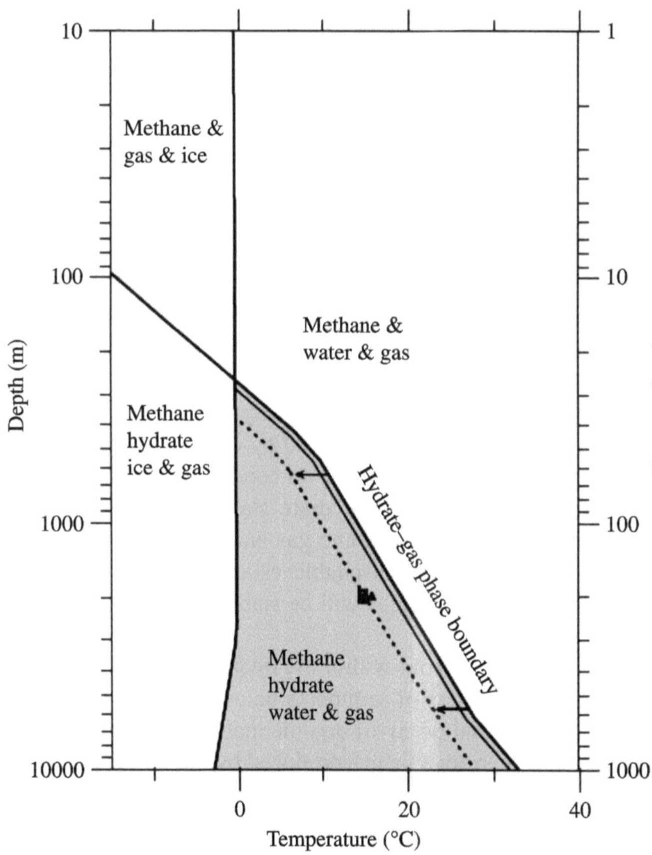

Fig. 4.8 Phase boundary diagram demonstrating the gas hydrate stability field (*gray*) (Kvenvolden and McDonald 1985)

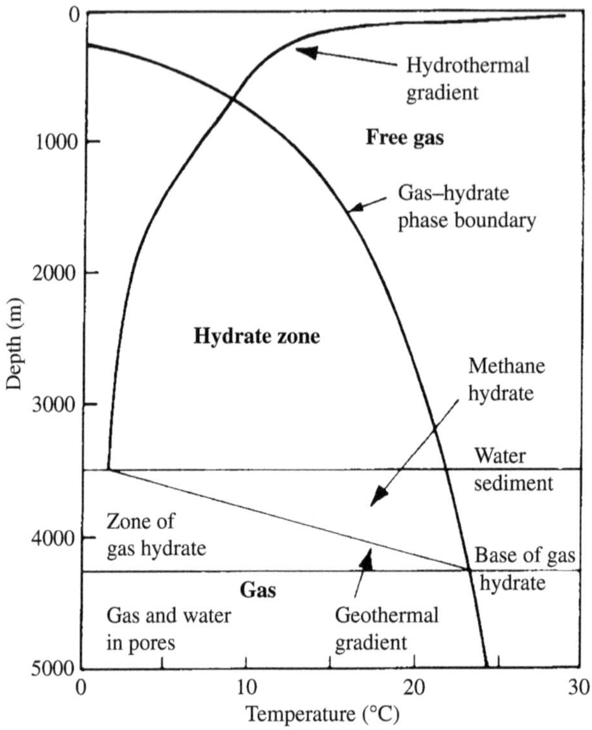

Fig. 4.9 A phase diagram of gas hydrate stability (Rastogi *et al.* 1999)

system. The directions of the hydrate–gas phase boundary shift toward stability at higher salinities are indicated by arrows. The thin line and the dashed line parallel to the freshwater hydrate–gas phase boundary indicate the seawater hydrate–gas phase boundary and the brine hydrate–gas phase boundary, respectively. Studies of deep-sea gas hydrates have been seriously hampered by difficulties in recovering gas hydrates because of their instability in sea-level conditions (Fig. 4.8).

Figure 4.9 shows the phase diagram of gas hydrate stability. The phase diagram, showing the boundary between free methane gas and methane hydrate for a pure water–methane system, provides a reasonable estimate of the pressure–temperature conditions under which natural gas will be stable on continental margins (Miles 1995).

Methane in hydrates is derived both from within the hydrate stability zone and more importantly from the great thickness of sediments below. Both shallow, biogenic gas derived mainly from bacterial decay of organic matter buried along with the sediment and deeper sourced, thermogenic gas produced by thermal cracking of higher density hydrocarbons have been found within hydrate deposits. Biogenic methane is usually dominant (Kvenvolden 1995). The requirement for sediments rich in organic carbon explains the suitability of continental margin sediments. Methane formed from either biogenic or thermogenic activity will tend to migrate upward, and be trapped as hydrate within the hydrate stability zone or as free gas beneath it. Once trapped as hydrate, the methane will tend to stay concentrated in the hydrate stability zone (Paull *et al.* 1991; Max and Lowrie 1996). Oceanic methane hydrate and associated gas deposits could not have formed without gas flow in the sediments and the development of a gas–fluid interface. Otherwise the methane would be dispersed near its source of production and there would be no significant concentrations of either gas or hydrate.

4.8 Facilities for Recovery of Methane from Methane Hydrate

Scientists are researching specific concerns about methane hydrate recovery and use which include drilling safety issues, potential influences on global climate change as methane is a potent greenhouse gas and the natural release of vast quantities from hydrate deposits would affect the global carbon cycle, cost-effective transportation of the gas to the surface, and the possible impact of hydrate removal on ocean floor stability.

There are four main possibilities for recovering of methane:

1. Add heat and raise the temperature to above that necessary for hydrate dissociation.
2. Depressurize the section by pumping, especially within the free gas below the BSR. The hydrate may then dissociate downward into the low-pressure gas layer. However, the latent heat of dissociation must still be provided.
3. Add antifreeze such as methanol; it may be possible to recover the methanol with the gas for reuse.
4. Replace the methane in the hydrate with carbon dioxide.

An intriguing possibility is to inject the unwanted greenhouse gas carbon dioxide into natural methane hydrate deposits, where it forms carbon dioxide hydrate in exchange for methane gas, which, in turn, is recovered. Carbon dioxide hydrate appears to be more energetically favorable than methane hydrate, so such a replacement should occur. This an attractive way to get rid of troublesome carbon dioxide and recover valuable methane.

Since methane gas could be recovered by the dissolution of methane hydrates, natural gas hydrates are a potential future energy supply. On the other hand, hydrates have been suggested as a natural gas storage medium (Ota *et al.* 1997) because of their high storage efficiency. Compared with LNG, hydrates have been proved to have the advantages of lower storage pressure and temperature at a lower cost (Gudmundsson and Borrehaug 1996).

In the last few years, many governments (including those of the USA, Canada, Russia, India, and Japan) have become very interested in possibilities of methane hydrates. Gas hydrates represent one of the world's largest untapped reservoirs of energy and, according to some estimates have the potential to meet global energy needs for the next 1,000 years.

Gas hydrates are of considerable interest because of their potential as an energy resource and because of their potential role in global climate change. From an energy resource point of view, the enormous amounts of methane hydrate under the ocean and beneath Arctic permafrost represent an estimated 53% of all fossil fuel reserves on Earth, about 10,000 gigatons (Lee and Holder 2001). The world-wide amounts of organic carbon bound in gas hydrates is conservatively estimated to total twice the amount of carbon to be found in all known fossil fuels on Earth (Satoh *et al.* 1996; Kvenvolden 1998; Collett 2002).

The difficulty in recovering this source of energy is that the fuel is in solid form and is not amenable to conventional gas and oil recovery techniques (Lee and Holder 2001). Proposed methods of recovery of gas from hydrates generally deal with dissociating or melting *in situ* gas hydrates by heating the reservoir beyond the temperature of hydrate formation, or decreasing the reservoir pressure below hydrate equilibrium. Models have been developed to evaluate production of natural gas from hydrates by both depressurization and heating methods (Goel *et al.* 2001).

4.8.1 Disposal of Impurities from Methane

In situations where natural gas and associated gas contain a lot of nitrogen, carbon dioxide, and hydrogen sulfide, hydrate technology can potentially be used to separate these gases from the source gas. This is because gas hydrates are thermodynamic equilibrium products. Mass transfer operations can be designed to carry out the separation and cleaning processes. In situations where saline water and brackish water need to be cleaned, gas hydrates can be produced and separated from the concentrated solution. This is because gas hydrates consume just water and gas, not other constituents such as dissolved salts and biological materials.

In situations where volatile organic compounds need to be recovered, for example, on oil tankers and receiving terminals, the hydrate-forming gases can be captured in the form of hydrate. The hydrate can be stored and then melted when the volatile organic compound gases can be used as fuel or blended with other hydrocarbons.

In situations where carbon dioxide disposal is needed, hydrate technology can be used to capture the gas in the form of a hydrate. Carbon dioxide hydrate can be transported by shuttle tankers and released at depth into the ocean. Because it is denser than seawater, the carbon dioxide hydrate will sink to the bottom. Provided the depth is greater than 250 m, the carbon dioxide hydrate should stay stable for practical purposes (Farkhondeh and Gheisi 2002).

4.8.2 Gas Hydrate Research and Development

Researchers have identified a need to better understand how the geological features in the permafrost regions and on continental margins control the occurrence and formation of methane hydrates. They have underscored the need to understand fundamental aspects (porosity, permeability, reservoir temperatures) of the geological framework that hosts the gas hydrate resource to improve assessment and exploration, to mitigate the hazard, and to enhance gas recovery.

Together with advances in research and development, economic viability will depend on the relative cost of conventional fuels, as well as other factors, such as pipelines and other infrastructure needed to deliver methane gas hydrate to market. Additionally, price volatility will likely affect the level of private sector investment in commercial production of gas hydrates. Table 4.1 shows international activities in gas hydrate research and development.

Gas hydrates in continental shelf sediments can become unstable either as a result of warming bottom water or as a result of a pressure drop due to a reduction in sea level (such as during an ice age). If these marine gas hydrates begin to rapidly disassociate into gas and water, then the methane trapped in the gas hydrates can be released to the atmosphere. Anthropogenically formed gas hydrates are another reason why these substances are of interest. Gas hydrates can spontaneously form in petroleum production equipment and pipelines associated with deepwater petroleum production and Arctic onshore petroleum production. These unwanted hydrates can clog equipment, preventing the optimum production of hydrocarbons.

The economic production of natural gas from oceanic hydrate deposits will not be realized using offshore drilling systems. The high cost of semisubmersible or tethered, deepwater, offshore drilling equipment is economically justified by the high production of dry gas over a long time. By definition, hydrated formations will produce wet gas at low pressure and low production rates. Fortunately hydrates are located relatively close to the seafloor compared with conventional gas reservoirs. Large numbers of low-cost wells could produce enough gas to recover costs. A workboat-based drilling system that includes an integrated gas production capability has been described.

A very modest amount of natural gas hydrate research and development has been performed to date. Most of it has been focused on gas industry operations, with the objective of finding better and/or cheaper means of ensuring that natural gas hydrates do not cause problems during the production, transportation, and distribution of conventionally sourced natural gas.

Table 4.1 International activities in gas hydrate research and development

Country	National gas hydrate programs active since
Japan	1995
India	1996
USA (2nd program)	1999
Germany	2000
South Korea	2001
China	2001

4.9 Properties of Seawater

Seawater is a mixture of 96.5% pure water and 3.5% other material, such as salts, dissolved gases, organic substances, and undissolved particles. The physical characteristics of seawater are determined by the physical properties, which are temperature, salinity, density, transparency, and ability to transmit light and sound. The most important physical factors for marine organisms are light, temperature, salinity, hydrostatic pressure (the weight of the water acting on a unit area), and acid–base balance. Physical parameters such as pressure, temperature, conductivity, and pH are measured through the water column. Salinity, density, and sound velocity are calculated from these measurements. Vertical profiles of temperature and salinity change locally. Vertical and horizontal distributions of parameters are also used for determining physical properties (Brown *et al.* 1997).

Two of the most important properties of seawater are the temperature and the salinity, because together they control its density, which is the major factor governing the vertical movement of ocean waters. The density of seawater decreases with increasing temperature and increases with increasing salinity and pressure.

The surface temperature of the sea depends on the insulation and determines the amount of heat radiated back into the atmosphere. Heat is also transferred across the surface of the sea by conduction and convection, and by the effects of evaporation. The average concentration of dissolved salts in the oceans, or the salinity, is about 35 g/kg.

Sound is a form of energy transmitted by rapid pressure changes in an elastic medium. Sound velocity is used for measurement of seawater depth, thickness, and characteristics of the earth's crust, and for detecting and investigating some organisms, especially fish. Sound velocity increases with increasing temperature and pressure, and decreases with increasing salinity (Geldiay and Kocabas 1998).

As on land, sunlight is essential for life in the sea. The depth to which light penetrates is also limited by the concentration and characteristics of particles in the water, and these particles scatter and absorb light.

Seawater density is a function of temperature and salinity. Seawater density (mass per unit volume) affects vertical circulation in the ocean. Small density differences caused by surface heating or cooling can produce strong vertical currents. Density can be measured directly by weighing a known volume of seawater. Also, instruments such as hydrometers and refractometers can be used. Density is typically expressed in grams per cubic centimeter. A typical seawater density is 1.0261 g/cm^3.

The densest water in the ocean is Antarctic bottom water. Its density is the result of low temperature and the formation of sea ice. As ice forms, fresh water is "frozen" out of the seawater, leaving behind salty dense water. Figure 4.10 shows an experimental setup for ice formation.

Temperature, salinity, and pressure (depth of the seawater column) play roles in determining the density of a seawater sample. As the temperature increases, the density decreases. The increase in temperature causes an increase in molecular activity, which translates into greater space between molecules. Table 4.2 shows

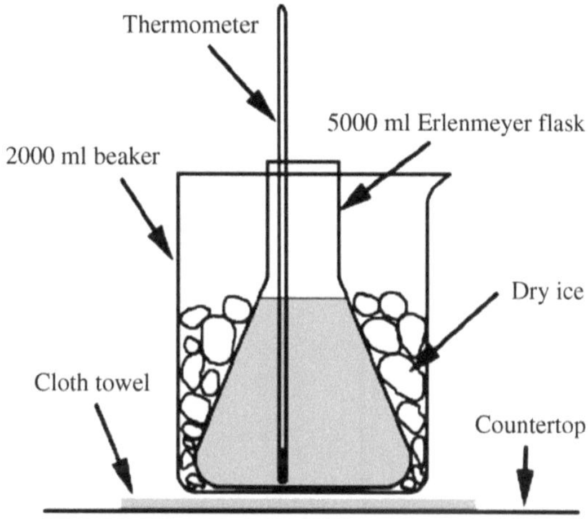

Fig. 4.10 Experimental setup for ice formation

a typical monthly distribution of temperature along the seacoast. Figure 4.11 shows average monthly temperature change in a seawater column.

As the salinity increases, the density increases. Salinity is usually measured as the total dissolved salts in a standard sample. As more salt is dissolved per unit volume, the density increases. As the pressure increases, the density increases. The deepest waters in the ocean are usually the densest owing to the weight of the overlying water.

Salinity is a measure of the amount of dissolved salts in seawater. The salt in seawater comes from the weathering of Earth's land surface. Table 4.3 shows various salt constituents of 1 kg of seawater. Two of the most important characteristics of ocean water are its temperature and its salinity. Together they help govern the density of seawater, which is the major factor controlling the ocean's vertical movements and layered circulation.

Table 4.2 Typical monthly distribution of temperature along the seacoast

Month	Air temperature (K)	Seawater temperature (K)
January	282.7	283.2
February	281.2	282.6
March	285.8	284.1
April	288.3	285.9
May	391.3	288.5
June	295.7	291.7
July	297.2	295.5
August	299.2	298.8
September	296.4	295.5
October	291.3	290.7
November	288.7	287.5
December	284.8	284.2

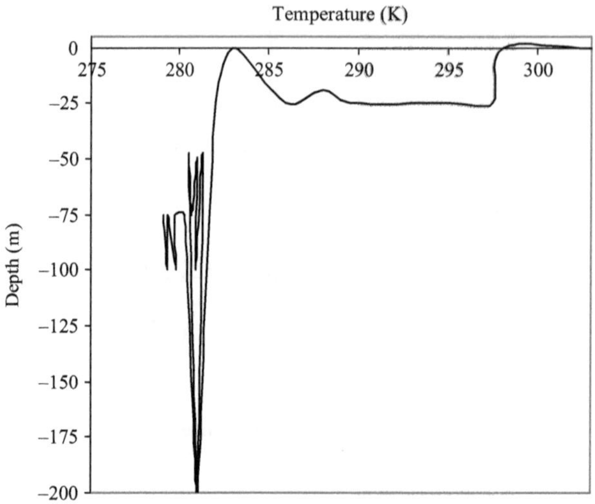

Fig. 4.11 Average monthly temperature change in a seawater column

Figure 4.12 shows a typical temperature–salinity diagram. In a temperature–salinity diagram, temperature is plotted along the vertical axis in degrees Celsius and salinity is plotted along the horizontal axis in parts per thousand or numerically equal practical salinity units. Seawater density, in grams per cubic centimeter, is shown in the diagram by curved lines of constant density. The value of each curved line appears immediately above each line. In the temperature–salinity diagram, each seawater sample is plotted as a dot at the point determined by its temperature and salinity. The temperature and salinity for the two surface seawater samples are labeled "A" and "B." If surface waters of the same density are brought together, they tend to mix. The temperature and the salinity of the resulting mixture are somewhere between the temperatures and salinities of the original waters prior to mixing. As shown by the temperature–salinity diagram, the density of seawater increases with increasing or decreasing temperature and with increasing

Table 4.3 Various salt constituents of 1 kg of seawater

Constituent	Mass (g)	Running total (g)
NaCl	28.014	28.014
$MgCl_2$	3.812	31.826
$MgSO_4$	1.752	33.578
$CaSO_4$	1.283	34.861
K_2SO_4	0.816	35.677
$CaCO_3$	0.122	35.799
KBr	0.101	35.900
$SrSO_4$	0.028	35.928
H_2BO_3	0.028	35.956

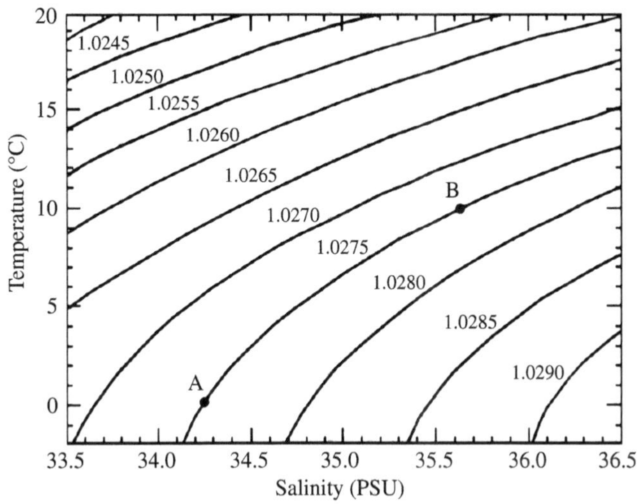

Fig. 4.12 Temperature–salinity diagram. *PSU* practical salinity units

or decreasing salinity. The temperature–salinity diagram is a simple but powerful tool used in studies of seawater density, mixing, and circulation.

A practical model for the effect of salinity on gas hydrate formation was investigated by Edmonds et al. (1996). The model represents the ionic components in water by a single salt pseudocomponent of the equation of state. The physical properties of the pseudocomponent were set by regressing them to experimental data for sodium chloride solutions. The freezing point and the maximum density are a function of salinity and temperature (Fig. 4.13).

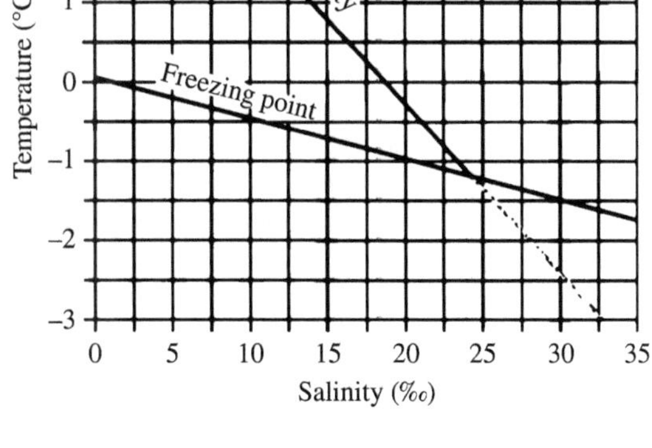

Fig. 4.13 Freezing point and maximum density as a function of salinity and temperature

As sodium chloride is usually the dominant component in produced water or seawater, other salts are handled on a sodium chloride equivalent basis, so only one salt pseudocomponent is needed for practical calculations. In practice, a hydrate inhibitor may be added to the water phase, so it is important that the inhibition model can give accurate predictions in the presence of saline solutions (Edmonds *et al.* 1996).

4.10 Summary

Gas hydrates are potentially one of the most important energy resources for the future. Methane gas hydrates are increasingly considered to be a potential energy resource. Enormous reserves of hydrates can be found under continental shelves and on land under permafrost.

Methane hydrates are widespread in sea sediments hundreds of meters below the seafloor along the outer continental margins and are also found in Arctic permafrost. Some deposits are close to the ocean floor and at water depths as shallow as 150 m, although at low latitudes they are generally only found below 500 m. The deposits can be 300–600 m thick and cover large horizontal areas. A nearby deposit nearly 500 km in length is found along the Blake Ridge off the coast of North Carolina at depths of 2,000–4,000 m.

Methane gas hydrates are of interest primarily for three reasons:

1. Gas from hydrate may be a new clean energy source. It is now recognized that there are huge amounts of natural gas, mainly methane, tied up in gas hydrates globally. Methane gas hydrates are a potential energy resource.
2. Natural gas hydrate may play a role in climate change. Methane is a strong greenhouse gas, so its escape to the atmosphere from natural gas hydrate could result in global warming.
3. There are important production problems. Gas hydrate is a hazard in conventional hydrocarbon exploration, from shallow gas release and from seafloor instability, especially in the Arctic and in deep water, where hydrate is stable.

Hydrates may affect the climate because when they are warmed or depressurized, they decompose and dissociate into water and methane gas, one of the greenhouse gases that warms the planet. Discharge of large amounts of methane into the atmosphere will cause global warming. Methane hydrates hold the danger of natural hazards associated with seafloor stability and release of methane to the ocean and atmosphere, and gas hydrates disturbed during drilling are a safety problem. Continental slope instability caused by hydrate decomposition is suggested as a trigger mechanism for underwater landslides and tsunami generation. If large volumes of methane are stored in marine reservoirs, they may significantly influence the sedimentary environment in which they occur.

Methane hydrates are located in the shallow submarine geosphere, which is a finely balanced system in equilibrium with all its components, such as sediment, pore water, fluid flows, pressure, temperature, overlying water, and hydrate. Removal of any one component of this equilibrium may destabilize the whole system, leading to irreparable damage. The destabilizing factors may be either natural perturbations or perturbations associated with exploitation. Studies have indicated that methane hydrates have the potential to affect global climate and the geological environment on a catastrophic scale.

Methane hydrates are common in sediments deposited on high-latitude continental shelves and at the slope and rise of continental margins with high bioproductivity. High biological production provides the organic matter buried in the sediment, which during early diagenesis and after exhausting oxygen, sulfate, and other electron acceptors eventually generates methane through fermentative decomposition and/or microbial carbonate reduction. The properties of sediment-hosted gas hydrates are strongly determined by the texture, structure, and permeability of the sediment and the mode of supply of methane.

References

Bakker J (1998) Improvements in clathrate modeling II: the H_2O-CO_2-CH_4-N_2-C_2H_6 fluid system. In: Henriet P, Mienert J (eds) Gas hydrates – relevance to world margin stability and climate change. Geological Society, London, pp 75–105

Bearat H, McKelvy M, Chizmeshya A, Nunez R, Carpenter R (2003) Investigation of the mechanisms that govern carbon dioxide sequestration via aqueous olivine mineral carbonation. In: Proceedings of the international technical conference on coal utilization and fuel system, vol 1, p 307

Blumenberg M, Seifert R, Michaelis W (2007) Aerobic methanotrophy in the oxic–anoxic transition zone of the Black Sea water column. Org Geochem 38:84–91

Blunier T (2000) Frozen methane escapes from the sea floor. Science 288:68–69

Blunier T, Chapellaz J, Schwander J, Stauffer B, Raynaud D (1995) Variations in methane concentration during the Holocene epoch. Nature 374:46–49

Boswell R, Collett TS (2006) "The gas hydrate resource pyramid," fire in the ice. Methane Hydrate R&D Program newsletter, pp 5–7. http://www.netl.doe.gov/technologies/oil-gas/FutureSupply/MethaneHydrates/newsletter/newsletter.htm

Brewer PG, Orr FM, Friederich G, Kvenvolden KA, Orange DL, McFarlane J, Kirkwood W (1997) Deep ocean field test of methane hydrate formation from a remotely operated vehicle. Geology 25:407–410

Brook EJ, Sowers T, Orchardo J (1996) Rapid variations in atmospheric methane concentrations during the past 110,000 years. Science 273:1087–1091

Brown E, Colling A, Park D, Philips J, Rothery D, Wright J (1997) Seawater: its composition, properties and behaviour. Open University, Milton Keynes

Bugge T, Befring S, Belderson RH, Eidvin T, Jansen E, Kenyon NH, Holtedahl H, Sejrup HP (1987) A giait three-stage submarine slide off Norway. Geol Mar Lett 7:191–198

Buffett BA, Zatsepina OY (2000) Formation of gas hydrate from dissolved gas in natural porous media. Mar Geol 164:69–77

Cicerone RJ, Oremland RS (1998) Biogeochemical aspects of atmospheric methane. Glob Biogeochem Cycles 2:229–327

Chagger HK, Kendall A, McDonald A, Pourkashanian M, Williams A (1998) Formation of dioxins and other semi-volatile organic compounds in biomass combustion. Appl Energy 60:101–114

Chapellaz J, Blunier T, Raynaud D, Barnola JM, Schwander J, Stauffer B (1993) Synchronous changes in atmospheric CH4 and Greenland climate between 40 and 8 kyr ago. Nature 366:443–445
Clennell MB, Hovland M, Booth JS, Henry P, Winters WJ (1999) Formation of natural gas hydrates in marine sediments. 1. Conceptual model of gas hydrate growth conditioned by host sediment properties. J Geophys Res 104(B10):22985–23003
Collett T (2000) Natural gas hydrate as a potential energy resource. In: Max M (ed) Natural Gas Hydrate in Oceanic and Permafrost Environment, p 123, Kluwer Publ., Livermore, CA
Collett TS (1993) Natural gas hydrates of the Prudhoe Bay and Kuparuk River area, North Slope, Alaska. AAPG Bull 77:793–812
Collett TS (1997) Gas hydrate resources of northern Alaska. Bull Can Pet Geol 45:317–338
Collett TS (2002) Energy resource potential of natural gas hydrates. AAPG Bull 86:1971–1992
Collect TS (2004) Alaska North Slope gas hydrate energy resources. USGS open file report no 1454
Collett TS, Bird KJ, Kvenvolden KA, Magoon LB (1989) Gas hydrates of Arctic Alaska. AAPG Bull 73:345–346
Crouch EM, Heilmann-Clausen C, Brinkhuis H, Morgans HEG, Rogers KM, Egger H, Schmitz B (2001) Global dinoflagellate event associated with the late Paleocene thermal maximum. Geology 29:315–318
Demirbas A (2005) Potential applications of renewable energy sources, biomass combustion problems in boiler power systems and combustion related environmental issues. Prog Energy Combust Sci 31:171–192
Demirbas A (2006) Hazardous emissions, global climate change and environmental precautions. Energy Sources Part B 1:75–84
Desa E (2001) Submarine methane hydrates potential fuel resource of the 21st century. Proc AP Akad Sci 5:101–114
Dickens GR (2000) Methane oxidation during the late Paleocene thermal maximum. Bull Soc Geol Fr 171:37–49
Dickens GR, O'Neil JR, Rea DK, Owen RM (1995) Dissociation of oceanic methane hydrate as a cause of the carbon isotope excursion at the end of the Paleocene. Paleoceanography 10:965–971
Dickens GR, Castillo MM, Walker JCG (1997) A blast of gas in the latest Paleocene: simulating first order effects of massive dissociation of oceanic methane hydrates. Geology 25:259–262
Durham WB, Kirby SH, Stern LA, Zhang W (2003) The strength and rheology of methane clathrate hydrate. J Geophys Res 108(B4):2182
Edmonds B, Moorwood R, Szczepanski R (1996) A practical model for the effect of salinity on gas hydrate formation. SPE paper 35569. Paper presented at the Society of Petroleum Engineers European production operations conference and exhibition, Stavanger, 16–17 April 1996
Eremeev VN, Konovalov SK, Romanov AS (1998)The distribution of oxygen and hydrogen sulfide in Black Sea waters during winter-spring period. Phys Oceonogr 9:259–272
Farkhondeh M, Gheisi AR (2002) An introduction to natural gas hydrate. Transportation. Methane gas hydrate report. Tehran University
Folger P (2008) Gas hydrates: resource and hazard. CRS report for Congress, November 26, 2008
Fujii T, Kawasaki M, Nakamizu M, Namikawa T, Ochiai K, Okui T, Tsuji Y (2005) Modes of occurrence and accumulation mechanism of methane hydrate – result of METI exploratory test wells "Tokai-oki to Kumano-nada". In: Fifth international conference on gas hydrates, Trondheim, Norway, pp 974–979
Geldiay R, Kocatas A (1998) Deniz biyolojisine giris. Ege University, Izmir
Goel N, Wiggins M, Shah S (2001) Analytical modeling of gas recovery from in situ hydrates dissociation. J Pet Sci Eng 29:115–127
Golomb D, Angelopoulos A (2001) A benign form of CO2 sequestration in the ocean. DOE NETL workshop on carbon sequestration science
Gornitz V, Fung I (1994) Potential distribution of methane hydrates in the world's oceans. Glob Biogeochem Cycles 8:335–347

Greinert J, Artemov Y, Egorov V, De Batist M, McGinnis D (2006) 1300-m-high rising bubbles from mud volcanoes at 2080 m in the Black Sea: hydroacoustic characteristics and temporal variability. Earth Plant Sci Lett 244:1–15

Gudmundsson JS, Borrehaug A (1996) In: The second international conference on natural gas hydrates, Toulouse, France, pp 415–422

Gupta AK (2004) Marine gas hydrates: their economic and environmental importance. Curr Sci 86:1198–1199

Haacke R.R, Westbrook GK, Riley MS (2008) Controls on the formation and stability of gas hydrate-related bottom-simulating reflectors (BSRs): a case study from the west Svalbard continental slope. J Geophys Res 113:1–12

Haq BU (1998) Natural gas hydrates: searching for the long-term climatic and slope-stability records. Geol Soc Lond Spec Publ 137:303–318

Hesselbo SP, Gröcke DR, Jenkyns HC, Bjerrum CJ, Farrimond P, Morgans Bell HS, Green OR (2000) Massive dissociation of gas hydrate during a Jurassic oceanic anoxic event. Nature 406:392–395

Hinrichs K-U, Hmelo LR, Sylva SP (2003) Molecular fossil record of elevated methane levels in late Pleistocene coastal waters. Science 299:1214–1217

Hornbach MJ, Holbrook WS, Gorman AR, Hackwith KL, Lizarralde D, Pecher I (2003) Direct seismic detection of methane hydrate on the Blake Ridge. Geophysics 68:92–100

Jakosky BM, Henderson BG, Mellon MT (1995). Chaotic obliquity and the nature of the Martian climate. J Geophys Res 100:1579–1584

Jean-Baptiste P, Ducroux R (2003) Energy policy and climate change. Energy Policy 31:155–166

Kalač P, Svoboda L, Havlíčková B (2004) Contents of detrimental metals mercury, cadmium and lead in wild growing edible mushrooms: a review. Energy Educ Sci Technol 13:31–38

Kannan N, Hong SH, Shim WJ, Yim UH (2007) A congener-specific survey for polychlorinated dibenzo-p-dioxins (PCDDs) and polychlorinated dibenzofurans (PCDFs) contamination in Masan Bay, Korea. Chemosphere 68:1613–1622

Kennett JP, Cannariato KG, Hendy IL, Behl RJ (2000) Carbon isotopic evidence for methane hydrate instability during Quaternary interstadials. Science 288:128–133

Kingston E, Clayton C, Priest J (2008) Gas hydrate growth morphologies and their effect on the stiffness and damping of a hydrate bearing sand. In: Proceedings of the 6th international conference on gas hydrates (ICGH 2008), Vancouver, British Columbia, Canada, July 6–10

Kubica K (1997) Distribution of PAH generated in domestic fuels boilers. In: Proceedings of the 9th international conference on coal science, Essen, September 7–12

Kurihara M, Ouchi H, Yoshihiro M, Hideo N, Yo O (2004) Assessment of gas productivity of natural methane hydrate using MH21 reservoir Simulator. The Research Consortium for Methane Hydrate Resources in Japan (MH21 Research Consortium), Tokyo, Japan

Kutz M (ed) (2007) Environmentally conscious alternative energy production. Wiley, Hoboken

Kvenvolden KA (1981) Methane hydrate – a major reservoir of carbon in the shallow geosphere? Chem Geol 71:41–51

Kvenvolden KA (1988a) Methane hydrates and global climate. Glob Biogeochem Cycles 2:221–229

Kvenvolden KA (1988b) Estimates of the methane content of worldwide gas-hydrate deposits, methane hydrates: resources in the near future? Paper presented at JNOC-TRC, Japan, 20–22 October

Kvenvolden KA (1988c) Methane hydrate – a major reservoir of carbon in the shallow geosphere? Chem Geol 71:41–51

Kvenvolden KA (1991) A review of Arctic gas hydrates as a source of methane in global change. In: Weller G, Wilson CL, Severin BAB (eds) International conference on the role of the polar regions in global change: proceedings of a conference held June 11–15, 1990 at the University of Alaska Fairbanks, Geophysical Institute and Center for Global Change and Arctic System Research. University of Alaska Fairbanks, pp 696–701

Kvenvolden K.A (1993) Gas hydrates – geological perspective and global change. Rev Geophys 31:173–187

Kvenvolden KA (1995) A review of the geochemistry of methane in natural gas hydrate. Org Geochem 23:997–1008
Kvenvolden KA (1998) A primer on the geological occurrence of gas hydrates. In Henriet J-P, Mienert J (eds) Gas hydrates – relevance to world margin stability and climate change. Geological Society, London, pp 9–30
Kvenvolden KA (1999) Potential effects of gas hydrate on human welfare. Proc Natl Acad Sci USA 96:3420–3426
Kvenvolden KA (2000) Natural gas hydrate: introduction and history of discovery In: Max MD (ed) Natural gas hydrate in oceanic and permafrost environments. Kluwer, Norwell, pp 9–16
Kvenvolden KA, McDonald TJ (1985) Gas hydrates of the Middle America Trench – deep sea drilling project Leg 84. In: von Huene R, Aubouin J et al (eds) Initial reports of the Deep Sea Drilling Project, vol 84. US Government Printing Office Washington, pp. 667–682
Kvenvolden KA, McMenamin MA (1980) Hydrates of natural gas: a review of their geologic occurrence. US Geol Surv Circ 825:1–11
Kvenvolden KA, Rogers BW (2005) Gaia's breath – global methane exhalations. Mar Pet Geol 22:579–590
Laberg JS, Vorren TO, Dowdeswell JA, Kenyon NH, Taylor J (2000) The Andoya slide and the Andoya canyon, north-eastern Norwegian-Greenland Sea. Mar Geol 162:259–275
Lavric ED, Konnov AA, De Ruyck J (2004) Dioxin levels in wood combustion – a review. Biomass Bioenergy 26:115–145
Lee S-Y, Holder GD (2001) Methane hydrates potential as a future energy source. Fuel Proc Technol 71:181–186
Lerche I, Bagirov E (1998) Guide to gas hydrate stability in various geological settings. Mar Pet Geol 15:427–438
Loveday JS, Nelmes RJ,Guthrie M, Belmonte SA, Allan DR, Klug DD, Tse JS, Handa YP (2001) Stable methane hydrate above 2 GPa and the source of Titan's atmospheric methane. Nature 410:661–663
Lunine JI, Stevenson DJ (1985) Thermodynamics of clathrate hydrate and low and high pressures with application to the outer solar system. Astrophys J Suppl Ser 58:493–531
MacDonald GJ (1990a) Role of methane clathrates in past and future climates. Clim Change 16:247–281
MacDonald GT (1990b) The future of methane as an energy resource. Annu Rev Energy 15:53–83
Mahajan D, Taylor CE, Mansoori GA (2007) An introduction to natural gas hydrate/clathrate: the major organic carbon reserve of the Earth. J Pet Sci Eng 56:1–8
Matsumoto R (1995) Causes of the d13C anomalies of carbonates and a new paradigm 'gas hydrate hypothesis'. J Geol Soc Jpn 101:902–924
Max MD, Lowrie A (1997) Oceanic methane hydrate development: reservoir character and extraction. In: Proceedings of the offshore technology conference, 5-8 May, Houston, Texas, pp 235–240
McIver RD (1981) Gas hydrates. in Meyer RF and Olson JC, eds., In: Long-term energy resources. Pitman, Boston, pp 713–726
Meyer RF (1981) in Meyer RF and Olson JC, eds., Speculations on oil and gas resources in small fields and unconventional deposits. In: Long-term energy resources. Pitman, Boston, pp 49–72
Michaelis W, Seifert R, Nauhaus K, Treude T, Thiel V, Blumenberg M, Knittel K, Gieseke A, Peterknecht K, Pape T, Boetius A, Aman A, Jørgensen BB, Widdel F, Peckmann J, Pimenov NV, Gulin M (2002) Microbial reefs in the Black Sea fueled by anaerobic oxidation of methane. Science 297:1013–1015
Miles PR (1995) Potential distribution of methane hydrate beneath the European continental margins. Geophys Res Lett 22:3179–3182
Milkov AV, Dzou L (2007) Geochemical evidence of secondary microbial methane from very slight biodegradation of undersaturated oil in a deep hot reservoir. Geology 35:455–458
Milkov AV, Etiope G (2005) Global methane emission through mud volcanoes and its past and present impact on the Earth's climate. Comment. Int J Earth Sci 94:490–492
Milkov AV, Sassen R (2001) Estimate of gas hydrate resource, northwestern Gulf of Mexico continental slope. Mar Geol 179:71–83

Milkov AV, Sassen R (2003) Preliminary assessment of resources and economic potential of individual gas hydrate accumulations in the Gulf of Mexico continental slope. Mar Pet Geol 20:111–128

Miller SL, Smythe WD (1970) Carbon dioxide hydrate and floods on Mars. Science 170:531–533

Murray JW (1991) Black Sea Oceanography. Results from the 1988 Black Sea expedition. Deep Sea Res 38:S655–S1266

Nisbet E, Piper DW (1998) Giant submarine slides. Nature 392:329–330

Nisbet EG (1990) The end of the ice age. Can J Earth Sci 27:148–157

Nixon MF, Grozic JLH (2006) A simple model for submarine slope stability analysis with gas hydrates. Nor J Geol 86:309–316

Ota M, Komatsu H, Murakami K, Sakamoto M (1997) Storage of natural gas hydrates as energy and water resources. In: International conference on fluid and thermal energy conversion, Yogyakarta, Indonesia, pp 413–417

Overmann J, Manske AK (2005) Anoxygenic phototrophic bacteria in the Black Sea. In: Neretin L (ed) Past and present water column anoxia. NATO ASI series. Springer, Berlin, chap II d.1

Overmann J, Manske AK (2006) Past and present water column anoxia. Springer, Dordrecht

Paull CK, Ussler W III, Dillon WP (1991) Is the extent of glaciation limited by marine gas hydrates? Geophys Res Lett 18:432–434

Piker L, Schmaljohann R, Imhoff JF (1998) Dissimilatory sulfate reduction and methane production in Gotland Deep sediments (Baltic Sea) during a transition period from oxic to anoxic bottom water (1993–1996). Aquat Microb Ecol 14:183–193

Pirkle JL, Kaufman RB, Brody DJ, Hickman T, Gunter EW, Paschal DC (1998) Exposure of the U.S. population to lead, 1991–1994. Environ Health Perspect 106:745–750

Popescu I, De Batist M, Lericolais G, Nouzé H, Poort J, Panin N, Versteeg W, Gillet H (2006) Multiple bottom-simulating reflections in the Black Sea: potential proxies of past climate conditions. Mar Geol 227:163–166

Rastogi A, Deka B, Bhattacharya GC, Ramprasad T, Kamesh Raju KA, Srinivas K, Murty GPS, Chaubey AK, Ramana MV, Subrahmanyan V, Sarma KVLNS, Desa M, Paropkari AL, Menezes AAA, Murthy VSN, Anthony MK, Subba Raju LV, Desa E, Veerayya M (1999) Gas hydrate stability zone thickness map of Indian deep offshore areas – a GIS based approach. In: Proceedings of the third international petroleum conference and exhibition Petrotech-99, New Delhi, pp 489–49

Raynaud D, Chapellaz J, Blunier T (1998) Ice-core record of atmospheric methane changes: relevance to climatic changes and possible gas hydrate sources. Geol Soc Lond Spec Publ 137:327–331

Reeburgh WS (1996) “Soft spots” in the global methane budget. In Lidstrom ME, Tabita FR (eds) Microbial growth on C1 compounds. Kluwer, Amsterdam, pp 334–342

Rodger PM (1990) Stability of gas hydrates. J Phys Chem 94:6080–6089

Sassen R, Sweet ST, Milkov AV, DeFreitas DA, Kennicutt MC (2001) Thermogenic vent gas and gas hydrate in the Gulf of Mexico slope: Is gas hydrate decomposition significant? Geology 29:107–110

Satoh M, Maekawa T, Okuda Y (1996) Estimation of amount of methane and resources of gas hydrates in the world and around Japan. J Geol Soc Jpn 102:959–971

Sauve S, McBride MB, Hendershot WH (1997) Speciation of lead in contaminated soils. Environ Pollut 98:149–155

Shirley K (2004) GOM gas hydrate opportunities explored – love ‘em or hate ‘em – they’re there. AAPG Explor (Jan 2004) 25:22–23

Sloan ED Jr (1998a) Physical/chemical properties of gas hydrates and application to world margin stability and climate change. In: Henriet J-P, Mienert J (eds) Gas hydrates – relevance to world margin stability and climate change. Geological Society, London, pp 31–50

Sloan ED Jr (1998b) Clathrate hydrates of natural gases, 2nd edn. Dekker, New York

Smith SL, Judge AS (1995) Estimates of methane hydrate volumes in the Beaufort-Mackenzie region, Northwest Territories. Curr Res-B; Geol Surv Can, March:81–88

Stern L, Circone S, Kirby S, Durham W (2001) J Phys Chem B 105:1756–1762
St. Louis VL, Kelly CA, Duchemin E, Rudd JWM, Rosenberg DM (2000) Reservoir surfaces as sources of greenhouse gases to the atmosphere: a global estimate. Bioscience 50:766–775
Subbotin OS, Ikeshojii T, Belosludov VR, Kudoh J, Belosludov RV, Kawazoe Y (2006) Local pressure and density distribution in methane hydrate –ice I*h* system. J Phys Conf Ser 29:206–209
Suess E (2002) Gashydrat – eine Verbindung aus Methan und Wasser. Nova Acta Leopoldina Neue Folge 85:123–146
Suess E, Bohrmann G, Rickert D, Kuhs WF, Torres ME, Thehu A, Linke P (2002) Properties and fabric or near-surface methane hydrates at Hydrate Ridge, Cassadia Margin. In: Proceedings of the fourth international conference on gas hydrates, Yokohama, May 19–23
Sultan N (2007) Comment on "Excess pore pressure resulting from methane hydrate dissociation in marine sediments: a theoretical approach" by Wenyue Xu and Leonid N. Germanovich. J Geophys Res Solid Earth 112:1–12
Sultan N, Cochonat P, Foucher J-P, Mienert J (2004) Effect of gas hydrates melting on seafloorslope instability. Mar Geol 213:379–401
Tame NW, Dlugogorski B, Kennedy EM (2007) Formation of polychlorinated dibenzo-p-dioxins and polychlorinated dibenzofurans (PCDD/F) in fires of arsenic-free treated wood: role of organic preservatives. Environ Sci Technol 41:6425–6432
Tester JW, Wood DO, Ferrari NA (1991) Economic policy in the face of global warming, energy and the environment in the 21st century. MIT Press, Cambridge
Thorpe RB, Law KS, Bekki S, Pyle JA, Nisbet EG (1996) Is methane driven deglaciation consistent with the ice core record? J Geophys Res 101:28:627–628, 635
Tohidi B, Anderson R, Clennell MB, Burgass RW, Biderkab AB (2001) Visual observation of gas-hydrate formation and dissociation in synthetic porous media by means of glass micromodels. Geology 29:867–870
Treude T, Orphan V, Knittel K, Gieseke A, House CH, Boetius A (2007) Consumption of methane and CO_2 by methanotrophic microbial mats from gas seeps of the anoxic Black Sea. Appl Environ Microbiol 73:2271–2283
Trofimuk AA, Cherskii NV, Tsaryov VP (1977) The role of continental glaciation and hydrate formation on petroleum occurrence. In: Meyer RF (ed) The future supply of nature-made petroleum and gas. New York, Pergamon Press, pp 919–926
US-OCSR (2008) Department of the Interior, Minerals Management Service, Resource Evaluation Division, preliminary evaluation of in-place gas hydrate resources: Gulf of Mexico outer continental shelf. OCS report MMS 2008-004 (Feb. 1, 2008). http://www.mms.gov/revaldiv/GasHydrateFiles/MMS2008-004.pdf
Waite WF, Winters WJ, Mason DH (2004) Methane hydrate formation in partially saturated Ottawa sand. Am Mineral 89:221–227
Weber A, Riess W, Wenzhoefer F, Jørgensen BB (2001) Sulfate reduction in Black Sea sediments: in situ and laboratory radiotracer measurements from the shelf to 2000 m depth. Deep Sea Res Part I Oceanogr Res Pap 48:2073–2096
Yefremova AG, Zhizhchenko BP (1974) Occurrence of crystal hydrates of gases in sediments of modern marine basins. Dokl Akad Nauk SSSR Earth Sci Sect 214:219–220
Zhang G, Rogers RE (2008) Ultra-stability of gas hydrates at 1 atm and 268.2 K. Chem Eng Sci 63:2066–2074

Chapter 5
Processes for Methane Production from Gas Hydrates

5.1 Introduction

The gases in the natural gas hydrate sediment are primarily methane molecules (Kvenvolden 1995). Methane hydrates are considered a major potential source of hydrocarbon energy and could be important in meeting natural gas demand in the future (Chi *et al.* 2006). Natural gas hydrates are a vast potential, though not presently commercial, source of additional natural gas. One of the most appealing aspects of this potential new gas source is that large deposits are located near the areas where demand for energy is expected to grow. Some countries, such as Japan, do not have indigenous oil or gas resources but do have nearby oceanic natural gas hydrate deposits. Means of economically and safely producing methane from gas hydrate deposits are not yet on the drawing board. The development of new, lower-cost technologies and approaches is required for economic production of methane gas from offshore hydrates. Low-cost production from gas hydrates will be limited by the rate of gas dissociation. Geothermal dissociation of gas hydrates provides the simplest, and lowest-cost, production technique. The dissociation reaction of gas hydrate is

$$CH_4\,6H_2O(s) \rightarrow CH_4(g) + 6H_2O(l) \qquad (5.1)$$

The production of natural gas from oceanic and permafrost sediments is currently being developed using methods such as depressurization, thermal stimulation, and injection of hydrate inhibitors (Moridis *et al.* 2004). The gas and water appear to flow in continuous pores of sediment during the hydrate decomposition, and thus it is necessary to investigate the pore network to understand the material flow in the sediment. It is important to understand the physical properties of sediment in investigations of structural properties, such as permeability, hydrate saturation, and sediment porosity, since these properties are essential to the development of natural gas production. The porosity is particularly important for material flow in sediment. Therefore, many reports have been published on the relationship between porosity and permeability (Noiriel *et al.* 2004; Quispe *et al.* 2005).

The difficulty in recovering this source of energy is that the fuel is in solid form and is not amenable to conventional gas and oil recovery techniques (Lee and Holder 2001). Proposed methods of gas recovery from hydrates generally deal with dissociating or melting *in situ* gas hydrates by heating the reservoir beyond the temperature of hydrate formation, or decreasing the reservoir pressure below hydrate equilibrium. Models have been developed to evaluate natural gas production from hydrates by both depressurization and heating methods.

The key problem in the production of methane from the hydrate layer is dissociation of *in situ* hydrates. Tapping the free methane below the gas hydrate zone is another possible method. The main steps involved in production from a prospective field are an exploration drilling program, fabrication and installation of the development/production platform, predrilling during construction of the platform, construction of a gathering system, production operations, and field abandonment. Three processes have been proposed for dissociation of methane hydrates: thermal stimulation, depressurization, and inhibitor injection. The obvious production approaches involve depressurization, heating, and their combinations.

The depressurization method involves lowering the pressure inside the well and encouraging the methane hydrate to dissociate. Its objective is to lower the pressure in the free-gas zone immediately beneath the hydrate stability zone, causing the hydrate at the base of the hydrate stability zone to decompose. Natural gas production from the dissociation of methane hydrate in a confined reservoir by a depressurizing down-hole well was studied. The case that the well pressure was kept constant was treated, and two different linearization schemes in an axisymmetric configuration were used in the analysis. It was shown that the gas production rate was a sensitive function of well pressure and reservoir temperature (Ahmadi *et al.* 2007).

The thermal stimulation method is applied to the hydrate stability zone to raise its temperature, causing the hydrate to decompose. In this method, a source of heat is provided directly in the form of injected steam or hot water or another heated liquid, or indirectly via electric or sonic means. This causes methane hydrate to decompose and generates methane gas. The methane gas mixes with the hot water and returns to the surface, where the gas and hot water are separated (Goel *et al.* 2001).

The chemical inhibition method seeks to displace the natural gas hydrate equilibrium condition beyond the hydrate stability zone's thermodynamic conditions through injection of a liquid inhibitor chemical adjacent to the hydrate. In this method, an inhibitor such as methanol is injected from the surface down to methane-hydrate-bearing layers.

The thermal stimulation method is quite expensive. The chemical inhibitor injection method is also expensive. The depressurization method may prove useful for efficient production of methane. Less carbon dioxide is produced by combustion of methane than by combustion of all other petroleum-derivate fuels, which makes methane attractive for use as an engine fuel in terms of the greenhouse effect. A very positive contribution of natural gas to minimization of environmental pollution must also be considered (Davies 2001; Demirbas 2006).

Of these three production methods, the depressurization combined with the thermal stimulation process appears to be the most practical for zones where free gas is trapped beneath the methane hydrates.

5.2 Depressurization Process

There are at least three means by which commercial production of natural gas hydrates might eventually be achieved, all of which alter the thermodynamic conditions in the hydrate stability zone such that the gas hydrate decomposes.

The first method is depressurization. Production is based on the depressurization-induced dissociation of the hydrates. Depressurization takes place in the section by pumping, especially within the free gas below the BSR. The hydrate may then dissociate downward into the low-pressure gas layer. However, the latent heat of dissociation must still be provided. The objective is to lower the pressure in the free-gas zone immediately beneath the hydrate stability zone, causing the hydrate at the base of the hydrate stability zone to decompose and the freed gas to move toward a well bore. Depressurization will therefore likely be the first production method tested outside the laboratory. In all cases, depressurization has involved the pore fluid; therefore, destabilization effects associated with changes in effective stress remain unknown. Multiple studies have explored the beneficial effects of combining heating and depressurization.

In the depressurization process, the hydrates are exposed to a low-pressure environment where they are unstable and decompose to methane and water. The heat energy for the process comes from the earth's interior. The methane released can then be recovered by conventional technology. The depressurization method as envisaged involves horizontal drilling in the underlying free-gas zone. As the free gas is removed, the overlying hydrate depressurizes and decomposes into free gas. Continuous removal of gas is expected to sustain this pressure-induced dissociation of the hydrate zone at its base. This method appears to be most suited to those deposits where widespread gas occurs in a closure below the hydrate cap (Max and Cruickshank 1999).

A variety of different mechanisms have been proposed for economically developing gas hydrates as an unconventional gas source (Sawyer *et al.* 2000; Goel *et al.* 2001). Thus far, the only method that has been successfully used to economically produce gas from gas hydrates is the depressurization method. This method is applicable only to hydrates that exist in polar regions beneath permafrost. It is applicable when a free-gas phase exists beneath the hydrate accumulation. Under such circumstances, production of the free-gas leg using conventional gas development techniques produces a pressure drop. This pressure drop causes the overlying hydrate to become unstable and to progressively disassociate into free gas and water, a process that adds gas to the underlying free-gas accumulation. Figure 5.1 shows gas production by the depressurization process.

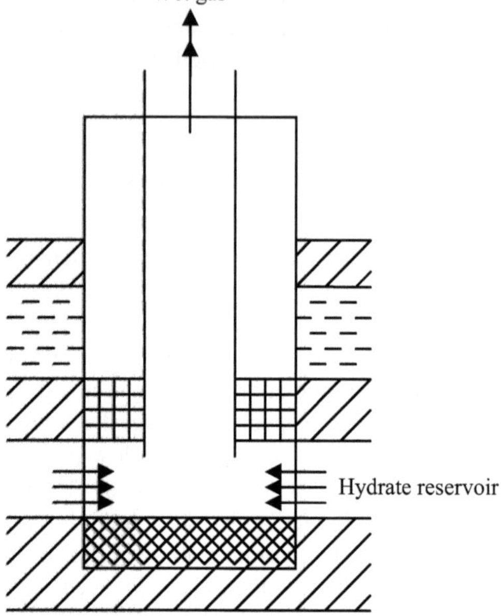

Fig. 5.1 Gas production by the depressurization process

For the depressurization process, the model behaves as a closed system with no boundaries. The following assumptions are considered (Dake *et al.* 1978; Ji *et al.* 2001; Khataniar *et al.* 2002):

1. Hydrate dissociation occurs as soon as the reservoir pressure drops below the dissociation pressure for the hydrate at the reservoir pressure. The gas flows immediately to the free-gas zone.
2. Hydrate decomposition is proportional to depressurization rate, and follows a first-order kinetic model.
3. Rock expansion and water expansion during gas production are negligible.
4. The model neglects heat transfer between the reservoir and the surroundings.
5. The reservoir is produced from a single well located at the center.

There are three important mechanisms involved in the depressurization of the gas hydrates: dissociation, heat transfer, including conduction and convection, and flow of fluids such as gas and water. The radial analytical model developed with decomposition kinetics by Goel *et al.* (2001) completely ignores the change in reservoir temperature due to the enthalpy of dissociation of hydrates and fluid flow. The model assumes a single-phase fluid flow where the dissociated water does not affect the flow, which is not quite true as water reduces the fluid transport within the system.

A one-dimensional model with heat transfer effects including conduction and convection was considered with a numerical solution that was generated by Ahmadi and Smith (2004) and it was mentioned that conductive heat flux plays a dominant role in supplying the heat for dissociation. It also highlights that the reservoir temperature affects the temperature gradient in the region near the front. This model gives the mass and heat balance at the dissociation front. A one-dimensional, three-phase fluid flow model with the effects of conduction and convection was considered to arrive at an analytical solution (Tsypkin 2000). The mass and energy conservation laws were used and natural gas was assumed to be a perfect gas. The kinetics was ignored. The conclusion of this work was that formation of ice leads to reduction in gas permeability and this affects the gas production volume. The work considered dissociation of gas hydrates coexisting with water and dissociation of hydrates at low temperatures.

5.3 Thermal Stimulation Process

The second method is thermal stimulation, in which a source of heat provided directly in the form of injected steam or hot water or another heated liquid, or indirectly via electric or sonic means, is applied to the hydrate stability zone to raise its temperature, causing the hydrate to decompose.

In the thermal stimulation process, heat energy can be released into the methane hydrate strata to dissociate the gas. This process has a favorable net energy balance, as the heat energy required for dissociation is about 6% of the energy contained in the liberated gas. In simple terms, steam or hot water can be pumped down a drill hole to dissociate the hydrate and release methane. The methane released can then be pumped to the surface of the seafloor through another drill hole (Desa 2001).

The direct approach could be accomplished in either of two modes: a frontal sweep similar to the steam floods that are routinely used to produce heavy oil, or by pumping hot liquid through a vertical fracture between an injection well and a production well. A major disadvantage of the thermal stimulation method is that a considerable portion of the applied energy (up to 75%) could be lost to non-hydrate-bearing strata (thief zones). A second major disadvantage is that the producing horizon must have good porosity, on the order of 15% or more, for the heat flooding to be effective. These drawbacks make the thermal stimulation method quite expensive. Figure 5.2 shows the gas production by thermal stimulation (heat injection) process.

Laboratory studies have been conducted on pure hydrate specimens (Circone *et al.* 2004, 2005; Stern *et al.* 2001, 2003) and in sediments with synthetic hydrate (Handa and Stupin 1992; Ogasawara *et al.* 2005; Sakamoto *et al.* 2005; Sung *et al.* 2002; Uchida *et al.* 2004; Yousif *et al.* 1991). Heating has been implemented by either injecting hot water or heating the chamber (Kamata *et al.* 2005; Kamath and Holder 1987; Ogasawara *et al.* 2005; Sakamoto *et al.* 2004, 2005; Ullerich *et al.* 1987).

In Japan, methane hydrate has considerable potential as a new energy resource. As a method for production of natural gas from the methane-hydrate-bearing layer, depressurization or depressurization with a well-wall heating process seems to be economically effective. The depressurization process decreases the system

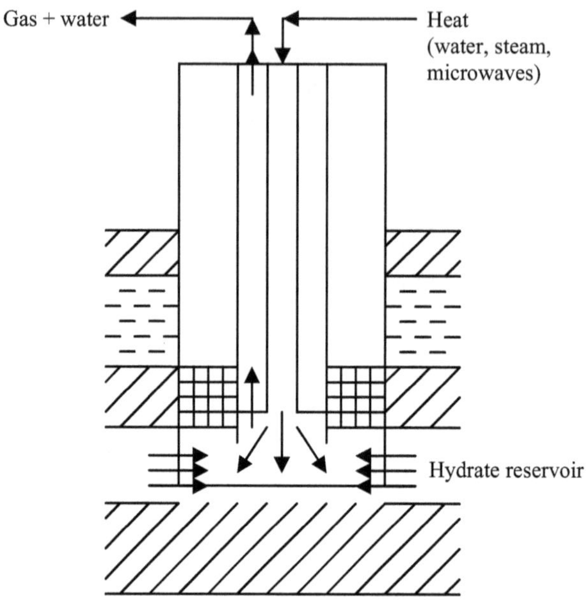

Fig. 5.2 Gas production by the thermal stimulation (heat injection) process

pressure below the pressure of hydrate formation at a specified temperature. The depressurization with a well-wall heating process is a combination of the thermal simulation method and the depressurization process. This process only heats the well wall at the hydrate-bearing layer; it is considered that the initial cost and the running cost is low. Therefore, a number of dissociation data are necessary for the assessment of the efficiency and to elucidate the decomposition process.

5.4 Chemical Inhibitor Injection Process

The third method is chemical inhibition, a concept similar to the chemical means presently used to inhibit the formation of water ice. This method seeks to displace the natural gas hydrate equilibrium condition beyond the hydrate stability zone's thermodynamic conditions through injection of a liquid inhibitor chemical adjacent to the hydrate. The chemical inhibitor injection method is also expensive, although less so than the thermal stimulation method, owing to the cost of the chemicals and the fact that it also requires good porosity. Figure 5.3 shows gas production by the chemical inhibitor injection process.

In the inhibitor injection process, an inhibitor such as methanol is injected into the gas hydrate zone. Chemical inhibitors shift the pressure–temperature equilibrium so that the hydrates are no longer stable at the *in situ* pressure–temperature condition, and hydrate dissociates at the contacted surface (Desa 2001).

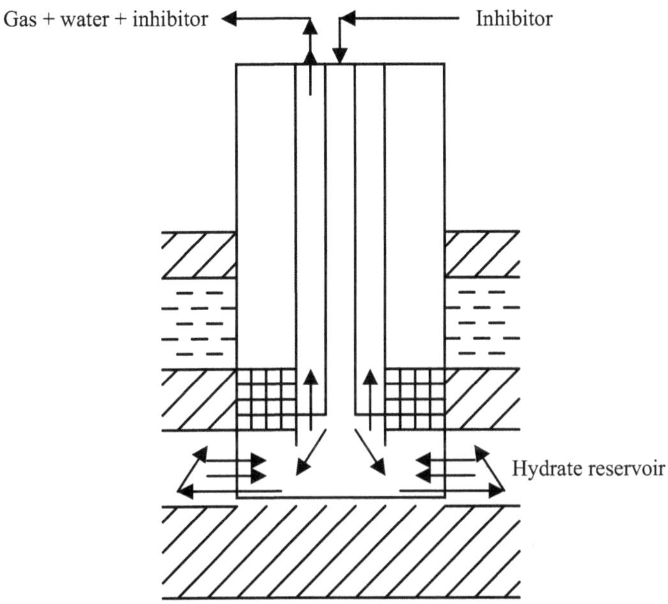

Fig. 5.3 Gas production by the chemical inhibitor injection process

An earlier work (Li *et al.* 2007) investigated the behavior of gas production from methane hydrate in porous sediment by injecting ethylene glycol solutions of different concentrations and at different injection rates in a one-dimensional experimental apparatus. The results suggest that the gas production process can be divided into four stages: (1) the initial injection, (2) the dilution of ethylene glycol, (3) the hydrate dissociation, and (4) gas output. Nevertheless, the water production rate stays nearly constant during the whole production process. The production efficiency is affected by both the ethylene glycol concentration and the ethylene glycol injection rate, and it reaches a maximum for an ethylene glycol concentration of 60 wt% (Li *et al.* 2007).

There are two approaches based on chemical concepts:

1. Chemical substitution: A very promising approach is the substitution of methane for carbon dioxide, thus recovering methane while sequestering carbon dioxide at the same time. The carbon dioxide can be brought into contact with the methane hydrate in the gas phase, in the liquid phase, or potentially dissolved in the circulating pore water (Ota *et al.* 2005a; Park *et al.* 2006). Raman spectroscopy, NMR, and MRI data provide insightful information about the evolution of the substitution (Graue *et al.* 2006; Gupta *et al.* 2005; Ota *et al.* 2005a, b; Sakamoto *et al.* 2005; Yoon *et al.* 2004). The optimal pressure–temperature operating conditions and underlying phenomena are reviewed in this chapter.
2. Chemical injection: Methanol, ethylene glycol, nitrogen, and salt brines are inhibitors that depress equilibrium conditions, so their injection prompts hy-

drate dissolution and methane production; their effect is intimately coupled with the imposed temperature difference (Chatterji and Griffith 1998; Graue *et al.* 2006; Haneda *et al.* 2005; Kamath *et al.* 1991; Kawamura *et al.* 2005; Ostergaard et al. 2005; Ota et al. 2005b; Sira *et al.* 1990; Sung *et al.* 2002, 2003; Yoon *et al.* 2004). Two solutions can be injected so that their exothermic reaction destabilizes the methane hydrate, liberating methane, and hindering its reformation by altering the fluid chemistry and lowering the phase transformation boundary (Chatterji and Griffith 1998). There is some evidence that nitrogen gas combined with heating is more effective than heating alone (Sakamoto *et al.* 2005).

Common inhibitors can be alcohols (methanol), glycols (ethylene glycol), and ionic salts. Several types of inhibitors have been tested with positive results, but it has been determined that glycols and alcohols are the most successful ones. The principle by which alcohols, glycols, and salts inhibit hydrates is the same. However, salts have some corrosion problems, and owing to low vapor pressures, they cannot vaporize. In the model reviewed here, methanol and ethylene glycol are examined and compared in many ways, such as chemical structure, physical properties, cost analysis, safety concentration limits, environmental considerations, and dehydration capacities.

Besides temperature and pressure conditions, the composition and the necessary amount of inhibitor must be determined. The inhibitor must be at or below its water dew point (i.e., must be water-saturated). In addition, dehydration can be used as an alternative. To find the amount of inhibitor needed to produce a unit amount of methane, first the amount of temperature depression and then the dissociation rate with respect to the weight percent concentration of inhibitors are calculated. Inhibition effects depend on both temperature and pressure, but pressure only slightly affects inhibition of hydrates so it is neglected in the modeling of inhibition. Again, it is assumed that there is continuous inhibitor injection to the system (Lederhos *et al.* 1996). Many aspects need to be considered during injection of inhibitors into gas hydrates, such as the shift in the equilibrium curve, the dissociation rate, and the properties of inhibitors.

The technology used with an ethylene glycol inhibitor results in lower cost, safer concentration limits, and a more environmentally attractive alternative to methanol for hydrate prevention in offshore gas lines. Ethylene glycol is recovered with a higher efficiency than methanol. Therefore, without recovery facilities, the large makeup volumes of methanol required to inhibit the high volume of gas would represent an extremely high operating cost. The unit prices of methanol and ethylene glycol are $ 0.84 and $ 4.75 per gallon, respectively. At first sight, it may seem that methanol used as an inhibitor costs less, but the recovery and recycling costs, which are operating costs, exceed the lower unit price of methanol. Furthermore, methanol poses greater safety risks in handling and storage than ethylene glycol. The flash points of methanol and ethylene glycol are 284 and 384 K, respectively, so methanol can be easily ignited.

5.5 Gas Hydrate Reservoirs

There are two gas hydrate reservoirs. They are Arctic hydrates and marine hydrates. Gas hydrates are found within and under permafrost in Arctic regions. They are also found within a few hundred meters of the seafloor on continental slopes and in deep seas and lakes. The reservoir architecture, technology needs, and eventual economic importance of hydrates in Arctic and marine environments may be very different. The commercial utilization of Arctic and marine gas hydrate resources is different.

5.5.1 Arctic Hydrates

Gas hydrates are found within and beneath permafrost on the North Slope of Alaska, in the Canadian Arctic, and in northern Siberia. The Arctic hydrates have the potential to become economically viable sources of natural gas. The best documented Alaskan accumulations are in the Prudhoe Bay–Kuparuk River area, which contains approximately 30 trillion standard cubic feet of natural gas, about twice the volume of conventional gas found in the Prudhoe Bay field (Collett 2002). The proximity to highly developed oilfield infrastructure makes the Prudhoe Bay–Kuparuk River accumulation particularly attractive. The main technology barrier is the lack of validated methods for economically viable production of natural gas from hydrate. An Arctic site capable of supporting multiyear field experiments would enable significant progress beyond the present state of knowledge.

Arctic hydrate reservoirs are potentially high quality sources of natural gas. The most important Arctic hydrate reservoirs accumulate in the Prudhoe Bay–Kuparuk River region on the North Slope of Alaska. Other reservoirs exist elsewhere on the North Slope of Alaska, in northern Canada, and in Siberia. Some important Arctic hydrate accumulations have good porosity and good gas saturation, and are predominantly found in coarse sands that have high intrinsic permeability. Overlying permafrost may provide a low-permeability barrier to gas leakage during extraction. These factors are favorable for production. One of the few fields in which gas production has been attributed to hydrates is the Messoyakha gas field in West Siberia.

A heavily monitored gas hydrate production experiment was carried out in 2002 at the Mallik site in the Mackenzie Delta, Northwest Territories, Canada (Dallimore and Collett 2005). During a period of 1 week, both depressurization and thermal stimulation were used to produce gas from hydrate. Although a technical success, the experiment was too brief to serve as a guide to reservoir-scale production.

In the absence of a gas pipeline, Arctic gas hydrate is a stranded resource. Generically there are three ways to destabilize gas hydrate to produce free gas from an Arctic gas hydrate reservoir. These are heating, depressurization, and chemical inhibition. Reservoir simulators have been developed to predict the efficacy of these techniques, singly or in combination (Masuda *et al.* 2002). Although consistent with each other, the simulators have yet to be validated by field test data (Moridis 2003).

There are several strong arguments in favor of a long-term gas hydrate production test facility in an Arctic area (AGHPW 2005):

- Because there are a number of ways of destabilizing hydrate, multiple tests will be required to identify the best technique or the best combination of techniques.
- Reservoir simulations suggest that these tests will need to be of long duration (1 year or more) to properly assess the success of a technique (Masuda *et al.* 2002; Moridis 2003).
- There are many advantages to drilling multiple wells from a single pad. Production tests in the Arctic can also serve as an early and relatively low cost indicator of how the potentially larger marine hydrate resources might be produced.

5.5.2 Marine Hydrates

Subsurface occurrences of natural gas hydrate can be classified into six types: (1) pore-space hydrate, (2) platy hydrate, (3) layered/massive hydrate, (4) disseminated hydrate, (5) nodule hydrate, and (6) vein/dyke hydrate. The anomalies of chloride contents in pore water, core temperature depression, core observation, as well as visible gas hydrates confirmed well-interconnected and highly saturated pore-space hydrates as an intergranular pore filling within sand layers within the methane hydrate stability zone. Hydrate saturations are higher than 60% throughout most hydrate-dominant sand layers and in some parts there is close to 100% pore saturation. Muddy sediments such as silts and clays are free of hydrate or contain low concentrations.

Figure 5.4 shows the sites where natural gas hydrate has been recovered or where its presence is inferred (Milkov 2004; Kvenvolden and Rogers 2005; Klauda and Sandler 2005). Subsea gas hydrates have been thought to comprise the preponderance of hydrate to be found in the geosphere. Moreover, they are to be found much closer to markets than are Arctic hydrates (Fig. 5.4). Promising accumulations are thought to exist off the east, west, and Gulf of Mexico coasts of the USA, as well as offshore Japan, India, China, and other important energy-consuming nations. Originally, the gas endowment was estimated to be 10,000 times larger than the global conventional gas endowment; however, most recent estimates are smaller, though large uncertainties persist.

Ocean Drilling Program investigations at a number of sites, including Blake Ridge offshore South Carolina, Hydrate Ridge offshore Oregon, and the Cascadia Margin offshore Vancouver, were funded in part by the National Science Foundation. With its abundant gas seeps in deep water, the Gulf of Mexico is probably the most promising marine gas hydrate province in US waters (Milkov and Sassen 2002). However, there has been no hydrate exploration activity there. A small number of hydrate-related wells were drilled in Keathley Canyon and Atwater Valley lease blocks in 2005, primarily to investigate drilling safety. This work was

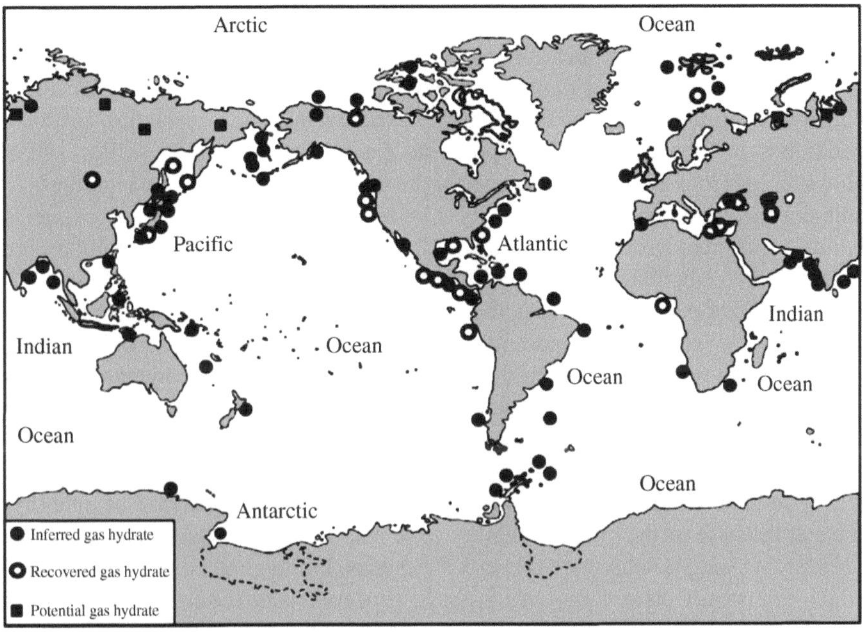

Fig. 5.4 Sites where natural gas hydrate has been recovered or where its presence is inferred

sponsored by US Department of Energy, and Chevron was the lead contractor. To achieve an assessment of marine gas hydrate resources, the US government would need to adopt the model used by the governments of Japan and India: an exploration and drilling campaign encompassing, for example, 20 sites dispersed over a significant geographic area.

5.6 Seafloor Drilling of the Hydrate Zone for Exploration and Production of Methane

The main cost here is only that of the pipeline used to transport the gas to the production platform. For subsea systems that do not produce to a fixed platform, a drilling template must be used that connects to a group of wells.

Drilling capabilities developed for conventional deepwater hydrocarbon exploration and production are presently sufficient for drilling and completing production in hydrate and associated gas deposits. Hydrate system deposits are always to be found relatively close to the seafloor. Hydrated sediments are expected in water depths between 500 and 2,000 m. Semisubmersible drilling systems with this depth capacity are currently available; however, the costs are extremely high (Brandt *et al.* 1998). These systems are capable of deep drilling for oil and gas with multilateral completions into reservoir compartments. Gas production is often avoided because of the cost of transport. To maintain reasonable gas costs, new technologies for recovery and transport are required.

Methane hydrate occurs in marine sediments in water depths greater than about 450 m on continental margins of open oceans. Below this, the temperature–pressure conditions in the sediment are appropriate for hydrate formation (Miles 1995). However, gas hydrates do not exist where the methane flux is too low, regardless of ambient thermodynamic conditions. Gas hydrates exist within seafloor sediments because sufficient gas and water are available to form hydrate in intergranular pore spaces. Oceanic hydrate system deposits occur mainly along continental slopes.

The hydrate economic zone is the combined hydrate, gas, and subjacent sediment zone for which it is important to characterize methane and the geotechnical properties that have a bearing on the gas recovery. It includes the hydrate stability zone and subjacent gas and pore fluid zones that are gas-rich. This is because the gas flux and transport to the hydrate economic zone, as well as methane interchange between hydrate and the gas phase, affects sediment properties and stability. In an area where sedimentation has continued over a long period of time, hydrate at the base of the hydrate stability zone may become unstable and dissociate because of rising geotherms. Where this happens, the hydrate conservation cycle, which is a steady-state, long-term process, conserves and concentrates the methane (Max and Lowrie 1996). Gas produced in or below sediment from dissociated hydrate will rise through buoyancy into the hydrate stability zone and tend to

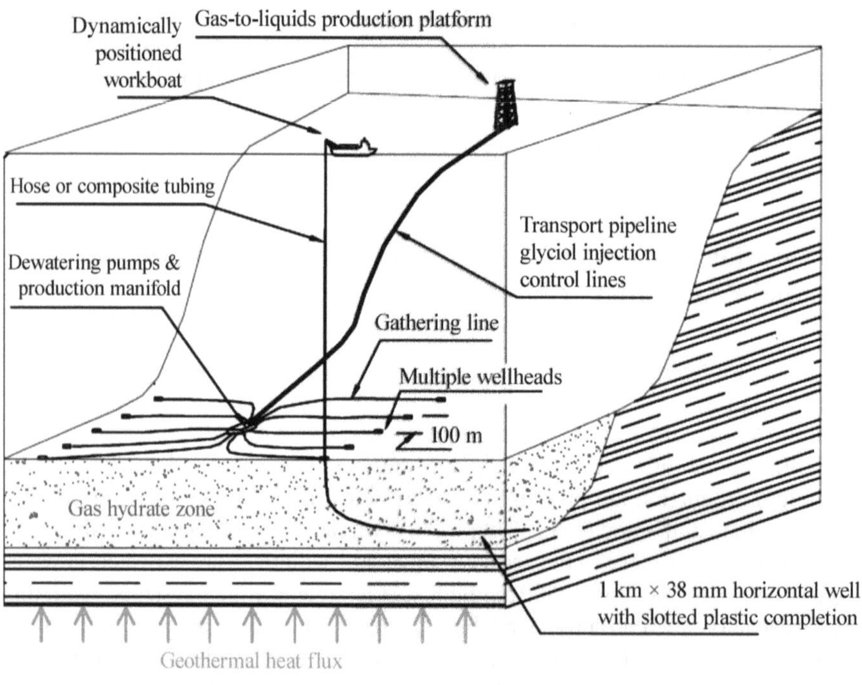

Fig. 5.5 Riserless drilling and production system for gas hydrates (Max and Chandra 1998)

again form gas hydrate. It is presently estimated that this economic zone is no more than 1.5–2 times the thickness of the hydrate stability zone. Below this zone, sediment compaction and geotechnical properties are more predictable and only gas generation occurs. The hydrate stability zone is relatively shallow, greatly limiting potential overpressures. Methane can normally penetrate the hydrate stability zone along faults (Dillon *et al.* 1997) and natural blowout of considerable volumes of methane has taken place.

In the case of a hydrate-culmination gas deposit, the drilling strategy can be either direct or indirect. In the first case, the gas reservoir is drilled directly through the hydrate and in the second case, the gas is drilled and tapped from the side, or possibly from below the gas closure, through the use of horizontal drilling techniques. Direct drilling into a normally pressurized gas trap does not usually result in undue safety or blowout problems because of the physical strength of the reservoir. Recent advances in drilling technology, which allow for considerable lateral and possible upward return drilling, might be used to avoid problems associated with direct drilling.

Avoiding technical problems associated with drilling directly down through the hydrate layer may be preferable to attempting to compensate for them. Indirect drilling, for instance, would penetrate the hydrate stability zone to the side of a bathymetric culmination gas trap, minimizing the likelihood of blowout. There is an additional advantage in having a long lateral hole in a gas reservoir: drawdown of gas would take place over a broad area through the reservoir rather than being localized near a vertical hole exposed to a shorter gas section.

A larger drawdown intersection may make it easier to maintain reservoir pressure and hydrate cap stability, as well as compensating for low or variable porosity and permeability.

Currently available drill string lengths are up to 8–10 km. In deep water, and to a lesser extent on continental slopes where terrigenous marine sediments are more common, the sediments are generally fine-grained silts and clays. As a result, the theoretical maximum drill string lengths may be realized in practice. Because the maximum water depths at which drilling can take place are no more than 5 km, the excess string length could be applied to curved and horizontal runs to indirectly tap hydrate-trapped gas.

Drilling within the hydrate economic zone, which extends no more than about 1–1.25 km below the seafloor mud line, can be carried out differently from conventional drilling. Ordinary drilling must penetrate much deeper below the seafloor and is likely to encounter a much wider variety of drilling conditions, including rocklike materials and substantially higher temperatures. To reduce the cost of completing the wells, a drill-in, telescopic casing system technique would be employed.

A workboat-based, riserless coiled tubing drilling system is proposed for this application. Composite coiled tubing would be used to drill in a horizontal production casing a few hundred meters beneath the seafloor. Large numbers of these low-cost wells could be tied together to support an offshore production facility and pipeline transport to shore. Figure 5.5 shows a riserless drilling and production system for gas hydrates.

Offshore operators have from time to time reported problems in drilling through gas hydrate zones. Drillers seeking conventional hydrocarbons have whenever possible purposely avoided drilling through natural gas hydrates because the process introduces two foreign sources of heat, friction and circulated drilling mud, that can cause dissociation of hydrates immediately adjacent to the borehole. When this is not avoidable, the hydrate stability zone is drilled and cased as fast as possible to minimize the risk of wall failure, perhaps leading to loss of the hole. Additionally, the free-gas zone beneath a hydrate cap can be overpressured, such that drilling into it without taking proper precautions can result in a blowout, just as is the case when conventional oil and gas drilling targets are involved. The Minerals Management Service has long maintained maps of the potential offshore natural gas hydrate occurrences to help ensure that this and the next category of risks are avoided or anticipated.

5.7 Possible Transportation Routes

If commercial production from oceanic natural gas hydrates is eventually established, there are at least three ways to transport the gas ashore: (1) by conventional pipeline; (2) by converting the gas hydrates to liquid middle distillates via the newly improved Fischer–Tropsch process and loading them onto a conventional tanker or barge; or (3) by reconverting the gas into solid hydrate and shipping it ashore in a close-to-conventional ship or barge.

Methane could be transported from the production site to the shore through submarine pipelines as is done for long-distance transportation of natural gas. However, submarine pipelines are expensive and the geological hazards of the continental slope make this option difficult. Other alternatives such as liquefying the gas on the ship or drilling platform can also be considered. One procedure involves burning part of the methane to get hydrogen and carbon monoxide, which can then be catalytically converted to liquid hydrocarbon. However, 35% of energy is lost in this process. Another method involves crystallizing the methane with water on the seafloor to obtain hydrate free from sediment. This pure hydrate then can be stored in zeppelin-shaped storage tanks and towed to a shallow water infrastructure where it can be safely decomposed into water and gas in a controlled environment (Desa 2001).

There are many transportation options for natural gas – CNG, LNG, pipelines, and gas-to-liquid (GTL) product transportation. Pipelines are generally attractive for a short distance from the shore owing to the low capital investments. However, pipeline transportation requires extensive gas processing, which will drive the production costs up, making pipeline transportation unfeasible for low production rates and high-moisture-containing gases.

LNG transportation requires a very high capital investment and for this reason should not be considered for the Hydrate Ridge site. Alternatively, CNG provides lower operating costs, but incurs high capital costs owing to the transportation

Fig. 5.6 Relationship of the economics of the yearly production rate and the distance to the market. *LNG* liquefied natural gas, *BCM* billion cubic meters, *CNG* compressed natural gas, *GTW* gas to wire, *NGH* natural gas hydrate, *GTL* gas to liquid

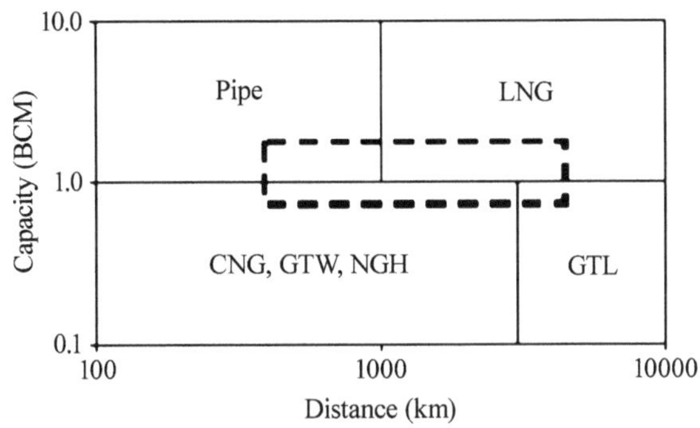

vessel. This makes CNG ideal for mid-sized reserves which can produce a large quantity of gas over the reserve lifetime to absorb the capital cost of the vessel. CNG is an attractive alternative if natural gas hydrate and GTL transportation proves to be a technical challenge or the production costs become too high. GTL transportation has the lowest capital investment for the processing equipment to produce the liquids, but is uneconomic at this point. If the cost of syngas production is lowered, GTL transportation will be very attractive and should be considered as the mode of transportation.

Figure 5.6 illustrates the relationship between annual production rate and distance from the market for the different modes of gas transportation (Gudmundsson 2003). As illustrated, LNG is not a feasible option for short distances or low production rates. Figure 5.6 presents GTL transportation as feasible for long distances for low production rates; this is because of the high capital cost for the GTL processors and CNG and natural gas hydrate become more economically feasible at shorter distances. However, this can be argued against and CNG for the Hydrate Ridge site is not as economically viable as hydrate shipping. The estimates for the cost used by Gudmundsson (2003) assume poor GTL conversions based on previous technologies. GTL technologies are becoming more economic and efficient and should not be ignored. On the basis of this analysis, hydrates, GTL transportation, and CNG are the more economic transportation modes, with CNG being a fallback if hydrates prove too technically challenging and GTL processes are too costly.

Overall, hydrate transportation seems to be an attractive transportation mode since extensive gas processing is unnecessary and the main costs are associated with the production of the hydrates. Since hydrate production can be performed at milder conditions than LNG and CNG production and GTL conversion, hydrate transportation is the most attractive route for delivering the recovered gas to the market place. On the basis of current research (Giavarini and Maccioni 2004), it could be feasible to transport methane hydrates at reasonable temperatures and pressures. It was found that the best storage conditions were 0.3 MPa and 269 K. Under these conditions there was slow dissociation of the hydrates, with complete dissociation occurring after 40 days, showing that methane hydrates may be a feasible commercial transport option over short periods of time.

Naturally occurring natural gas hydrates present both mechanical and chemical risks. Normal drilling can generate enough downhole heat to decompose surrounding hydrates, possibly resulting in loss of the well or in loss of well control and conceivably an ensuing loss of foundation support. Although large volumes of oceanic natural gas hydrate deposits are known to have decomposed in the past without human influence, information on their role in the global carbon cycle and global climate change is limited. It is clear that the release of large quantities of methane into the atmosphere, for whatever reason, would substantially increase its greenhouse capability since methane is 21 times more potent a greenhouse gas than is carbon dioxide.

5.8 Economic and Political Issues

The economic production of natural gas from oceanic hydrate deposits will require new offshore drilling systems and methods. Also, the product of hydrate dissociation may be relatively low pressure, wet gas, especially where the excess pressure produced by hydrate dissociation can be equilibrated rapidly through high-porosity sediments. Natural gas derived from hydrate may result in low local production rates. The low production rate requires low-cost drilling methods to ensure economic viability. Recovering methane and economically transporting it pose a challenge to technologists and scientists. Ideas have been conceptualized and research has been initiated to address these challenges.

Methane hydrate will be commercially exploited only when the price of petroleum oil and conventional gas rises substantially. The apparent abundance of conventional hydrocarbon deposits and their relatively low prices are inhibiting research into various aspects of gas hydrate. Research and development activities in this field need to be sustained. Hydrate recovery will in all probability involve forced dissociation, which will involve significant demand for heat. Supplying and managing this heat and maintaining an artificial thermodynamic balance that allows the controlled dissociation of hydrate and the safe recovery of methane will probably prove the key to commercialization.

On the basis of calculations, depressurization was shown to be the most promising technique for class 1 type reservoirs (Moridis 2003). Depressurization has also been quoted by many researchers as the most economically viable option (Makogon 1997; Pooladi-Darvish 2004).

Methanol is approximately 3 times less expensive than ethylene glycol. One must pay particular attention to the amount of methanol necessary to treat the inlet gas. With increasing gas flow rates, the ethylene glycol injection process typically becomes a more viable option because the inhibitor is regenerated. The increased cost of utilizing methanol injection to treat larger gas volumes can be directly associated with the raw material makeup cost.

Most methane hydrate deposit locations were discovered serendipitously when scientists looked anew at existing seismic data. These initial efforts did not require much expenditure. As the number of finds increased, knowledge about various aspects of oceanic hydrates also increased and scientists began realizing their immense potential as a fuel resource. The challenges of hydrate exploration and production and its probable impact on the global climate and the geological environment became clearer. Policymakers are gradually recognizing the long-term potential of marine methane hydrates as well. It appears, however, that the complexities and challenges for exploration of methane hydrates and their production from the hostile and difficult marine environment require considerable focused research and development efforts in various fields, for which adequate financial support is lacking. This is perhaps due to the perception that methane hydrate exploitation will be economically viable only when the price of conventional hydrocarbon and other fuels rises substantially.

On the other hand, many hydrate deposits are located on continental slopes not far from major markets in industrialized countries. Countries that have strong economic bases, or are witnessing high industrial growth rates, but have low energy resource potential, could potentially become energy-independent, an event that would affect international affairs, foreign policy, and other interrelations. The repercussions would extend to world trade, regional power equations, and the foreign currency balance of existing major importers when gas hydrate begins to be exploited. The realization that such a situation could come about has recently generated some interest in the field of gas hydrate research in many countries. It is expected that the prospect of energy self-reliance will catalyze some of these countries to initiate harvesting methane hydrates as soon as scientists and technologists come forward with dependable, safe, and cost-effective mechanisms to explore and exploit this resource.

Lack of suitable production technology was a major impediment in exploitation of this resource. However, the past 5 years has witnessed a dramatic improvement in drilling technologies for oil and gas in deepwater areas, where hydrate deposits occur. There has also been a distinct reduction in deepwater development costs. All these are positive factors for hydrate exploration and development. Much of the engineering required to exploit these deposits can be achieved by suitably adopting proven technology currently used in connection with exploitation of deepwater oil and gas reserves.

5.9 Summary

Three processes have been proposed for dissociation of methane hydrates: thermal stimulation, depressurization, and inhibitor injection. The obvious production approaches involve depressurization, heating, and their combinations. The depressurization method involves lowering the pressure inside the well and encouraging the methane hydrate to dissociate. The chemical inhibition method seeks to displace the natural gas hydrate equilibrium condition beyond the hydrate stability zone's thermodynamic conditions through injection of a liquid inhibitor chemical adjacent to the hydrate. Of these three production methods, the depressurization combined with the thermal stimulation process appears to be the most practical for zones where free gas is trapped beneath the methane hydrates.

There are two gas hydrate reservoirs. They are Arctic hydrates and marine hydrates. Gas hydrates are found within and under permafrost in Arctic regions. They are also found within a few hundred meters of the seafloor on continental slopes and in deep seas and lakes.

The main cost here is only that of the pipeline used to transport the gas to the production platform. For subsea systems that do not produce to a fixed platform, a drilling template must be used that connects to a group of wells. Transportation of methane from the production site to the shore could be through submarine pipelines as is done for long-distance transportation of natural gas. However, submarine pipelines are expensive and the geological hazards of the continental slope make this option difficult.

The economic production of natural gas from oceanic hydrate deposits will require new offshore drilling systems and methods. Recovering methane and economically transporting it pose a challenge to technologists and scientists. Ideas have been conceptualized and research has been initiated to address these challenges.

On the basis of calculations, depressurization was shown to be the most promising technique for class 1 type reservoirs. Depressurization has also been quoted by many researchers as the most economically viable option. Methanol is approximately 3 times less expensive than ethylene glycol. One must pay particular attention to the amount of methanol necessary to treat the inlet gas. With increasing gas flow rates, the ethylene glycol injection process typically becomes a more viable option because the inhibitor is regenerated. The increased cost of utilizing methanol injection to treat larger gas volumes can be directly associated with the raw material makeup cost.

References

AGHPW (2005) Alaska Gas Hydrate Planning Workshop (AGHPW), proceedings, sponsored by Alaska Department of Natural Resources and United States Geological Survey, Anchorage, Alaska, 17–18 August

Ahmadi G, Smith, DH (2004) Numerical solution for natural gas production from methane hydrate dissociation. J Pet Sci Eng 41:269–285

Ahmadi G, Ji C, Smith DH (2007) Production of natural gas from methane hydrate by a constant downhole pressure well. Energy Convers Manage 48:2053–2068

Brandt W, Dang AS, Magne E, Crowley D, Houston K, Rennie A, Hodder M, Stringer R, Juiniti R, Ohara S, Rushton S (1998) Deepening the search for offshore hydrocarbons. Oilfield Rev Spring 2–21

Chatterji J, Griffith JE (1998) Methods of decomposing gas hydrates. US Patent 5,713,416

Chi W-C, Reed DL, Tsai C-C (2006) Gas hydrate stability zone in offshore southern Taiwan. Terr Atmos Ocean Sci 17:829–843

Circone S, Stern LA, Kirby SH (2004) The role of water in gas hydrate dissociation. J Phys Chem B 108:5747–5755

Circone S, Kirby SH, Stern LA (2005) Thermal regulation of methane hydrate dissociation: Implications for gas production models. Energy Fuels 19:2357–2363

Collett TS (2002) Energy resource potential of natural gas hydrates. AAPG Bull 86:1971–1992

Dake LP (1978) Fundamentals of reservoir engineering. Elsevier, New York

Dallimore SR, Collett TS (2005) Scientific results from the Mallik 2002 gas hydrate production research well, Mackenzie Delta, Northwest Territories, Canada. Bulletin 585. Geological Survey of Canada, Ottawa

Davies P (2001) The new challenge of natural gas. Paper presented at OPEC and the global energy balance: towards a sustainable future, Vienna, 28 September 2001

Demirbas A (2006) The importance of natural gas in the world. Energy Sources Part B 1:413–420

Desa E (2001) Submarine methane hydrates potential fuel resource of the 21st century. Proc AP Akad Sci 5:101–114

Dillon WP, Holbrook WS, Drury R, Gettrust J, Hutchinson D, Booth J, Taylor M (1997) Faulting of gas-hydrate-bearing marine sediments – contribution to permeability. Paper 8296. In: Offshore technology conference Houston, Texas, 5–8 May 1997, pp 201–209

Giavarini C, Maccioni F (2004) Self-preservation at low pressures of methane hydrates with various gas contents. Ind Eng Chem Res 43:6616–6621

Goel N, Wiggins M, Shah S (2001) Analytical modeling of gas recovery from in situ hydrates dissociation. J Pet Sci Eng 29:115–127

Graue A, Kvamme B, Baldwin BA, Stevens J, Howard J, Ersland G, Husebo J, Zomes DR (2006) Magnetic resonance imaging of methane – carbon dioxide hydrate reactions in sandstone pores. In: 2006 SPE annual technical conference and exhibition, pp 24–27

Gudmundsson H (2003) Making concepts matter: sustainable mobility and indicator systems in transport policy. Int Soc Sci J 55:199–217

Gupta A, Dec SF, Koh CA, Sloan ED (2005) NMR and remain investigations of the methane hydrate dissociation mechanism. In: Proceedings of the fifth international conference on gas hydrates, June 12–16, Trondheim, Norway

Handa YP, Stupin D (1992) Thermodynamic properties and dissociation characteristics of methane and propane hydrates in 70-A-radius silica gel pores. J Phys Chem 96:8599–8603

Haneda H, Sakamoto Y, Kawamura T, Komai T (2005) Experimental study on dissociation behavior of methane hydrate by air. In: Proceedings of the fifth international conference on gas hydrates, June 12–16, Trondheim, Norway

Ji C, Ahamadi G, Smith DH (2001) Natural gas production from hydrate decomposition by depressurization. Chem Eng Sci 56:5801–5814

Kamata Y, Ebinuma T, Ota M, Minagawa H, Narita H, Masuda Y, Konno Y (2005) Decomposition experiment of methane hydrate sediment by thermal recovery method. In: Proceedings of the fifth international conference on gas hydrates, June 12–16, Trondheim, Norway

Kamath VA, Holder GD (1987) Dissociation heat transfer characteristics of methane hydrates. AIChE J 33:347–350

Kamath VA, Mutalik PN, Sira JH, Patil SL (1991) Experimental study of brine injection and depressurization methods for dissociation of gas hydrates. SPE Form Eval 6:477–484

Kawamura T, Yamamoto Y, Ohtake M, Sakamoto Y, Komai T, Haneda H (2005) Experimental study on dissociation of hydrate core sample accelerated by thermodynamic inhibitors for gas recovery from natural gas hydrate. In: Proceedings of the fifth international conference on gas hydrates, June 12–16, Trondheim, Norway

Khataniar S, Kamath VA, Omenihu SD, Patil SL, Dandekar AY (2002) Modeling and economic analysis of gas production from hydrates by depressurization method. Can J Chem Eng 80:135–143

Klauda JB, Sandler SI (2005) Global distribution of methane hydrate in ocean sediment. Energy Fuels 19:459–470

Kvenvolden KA, Rogers BW (2005) Gaia's breath – global methane exhalations. Mar Pet Geol 22:579–590

Lederhos JP, Long JP, Sum A, Christiansen RL, Sloan ED Jr (1996) Effective kinetic inhibitors for natural gas hydrates. Chem Eng Sci 51:1221–1229

Lee S-Y, Holder GD (2001) Methane hydrates potential as a future energy source. Fuel Proc Technol 71:181–186

Li G, Li X-S, Tang L-G, Zhang Y (2007) Experimental investigation of production behavior of methane hydrate under ethylene glycol injection in unconsolidated sediment. Energy Fuels 21:3388–3393

Makagon YF (1997) Hydrates of hydrocarbons. PennWell Books, Tulsa

Masuda Y, Kurihara M, Ohuchi H, Sato T (2002) A field-scale simulation study on gas productivity of formations containing gas hydrates. In: Proceedings of the fourth international conference on gas hydrates, Yokohama, Japan, 19–23 May

Max MD, Chandra K (1998) The dynamic oceanic hydrate system: production constraints and strategies. In: Proceedings of the offshore technology conference, 4–7 May 1998, Houston, Texas, pp 217–226

Max MD, Cruickshank MJ (1999) Extraction of methane from oceanic hydrate system deposits. In: Proceedings of the offshore technology conference, 3–6 May 1999, Houston, Texas, pp 1–8

Max MD, Lowrie A (1997) Oceanic methane hydrate development: Reservoir character and extraction. In: Proceedings of the offshore technology conference, 5–8 May, Houston, Texas, pp 235–240

Miles PR (1995) Potential distribution of methane hydrate beneath the European continental margins. Geophys Res Lett 22:3179–3182

Milkov AV (2004) Global estimates of hydrate-bound gas in marine sediments: how much is really out there? Earth Sci Rev 66:183–197

Milkov AV, Sassen R (2002) Economic geology of offshore gas hydrate accumulations and provinces. Mar Pet Geol 19:1–11

Moridis GJ (2003) Numerical studies of gas production from methane hydrates. Soc Pet Eng J Dec 359–370

Moridis GJ, Collett T, Dallimore S, Satoh T, Hancock S, Weatherhill B (2004) Numerical studies of gas production from several methane hydrate zones at the Mallik site, Mackenzie Delta, Canada. J Pet Sci Eng 43:219–239

Noiriel C, Gouze P, Bernard D (2004) Investigation of porosity and permeability effects from microstructure changes during limestone dissolution. Geophys Res Lett 31:L24603

Ogasawara K, Yamasaki A, Kiyono F, Kato C, Kawamura M (2005) Development of new apparatus for measuring dissociation rate of a methane hydrate under flow condition of water. In: Proceedings of the fifth international conference on gas hydrates, June 12–16, Trondheim, Norway

Ostergaard KK, Masoudi R, Tohidi B, Danesh A, Todd AC (2005) A general correlation for predicting the suppression of hydrate dissociation temperature in the presence of thermodynamic inhibitors. J Pet Sci Eng 48:70–80

Ota M, Abe Y, Watanabe M, Smith RL Jr, Inomata H (2005a) Methane recovery from methane hydrate using pressurized CO_2. Fluid Phase Equilib 228/229:553–559

Ota M, Morohashi K, Abe Y, Watanabe M, Smith RL Jr, Inomata H (2005b) Replacement of CH_4 in the hydrate by use of liquid CO_2. Energy Convers Manage 46:1680–1691

Quispe JR, Rozas R, Toledo PG (2005) Permeability-porosity relationship from a geometrical model of shrinking and lattice-Boltzmann and Monte Carlo simulations of flow in two-dimensional pore networks. Chem Eng J 111:225–236

Park YJ, Kim DY, Lee JW, Huh DG, Park KP, Lee JY, Lee H (2006) Sequestering carbon dioxide into complex structures of naturally occurring gas hydrates. Proc Natl Acad Sci USA 103:12690–12694

Pooladi-Darvish M (2004) Gas production from hydrate reservoirs and its modeling. Soc Pet Eng 56:65–71

Sakamoto Y, Komai T, Kawabe Y, Tenma N, Yamaguchi T (2004) Gas hydrate extraction from marine sediments by heat stimulation method. Paper presented at the international offshore and polar engineering conference Toulon, 23–28 May

Sakamoto Y, Komai T, Kawamura T, Tenma N, Yamaguchi T (2005) Field scale simulation for effect of permeability distribution and anisotropy on dissociation and gas production behavior in marine sediments with gas hydrate. Seoul, South Korea. International Society of Offshore and Polar Engineers, Cupertino, pp 386–391

Sawyer W, Boyer C, Franz J, Yost A (2000) Comparative assessment of natural gas hydrate production models. SPE Pap 62513

Sira JH, Patil SL, Kamata VA (1990) Study of hydrate dissociation by methanol and glycol injection. Soc Pet Eng 977–984

Stern LA, Circone S, Kirby SH, Durham WB (2001) Preservation of methane hydrate at 1 atm. Am Chem Soc 15:499–501

Stern LA, Circone S, Kirby SH, Durham WB (2003) Temperature, pressure, and compositional effects on anomalous or "self" preservation of gas hydrates. Can J Phys 81:271–283

Sung W, Lee H, Lee H, Lee C (2002) Numerical study for production performances of a methane hydrate reservoir stimulated by inhibitor injection. Energy Sources 24:499–512

Sung W, Lee H, Kim S, Kang H (2003) Experimental investigation of production behaviors of methane hydrate saturated in porous rock. Energy Sources 25:845–856

Tsypkin GG (2000) Mathematical models of gas hydrates dissociation in porous media. Ann N Y Acad Sci 912:428–436

Uchida T, Takeya S, Chuvilin EM, Ohmura R, Nagao J, Yakushev VS, Istomin VA, Minagawa H, Ebinuma T, Narita H (2004) Decomposition of methane hydrates in sand, sandstone, clays, and glass beads. J Geophys Res 109:B05206

Ullerich JW, Selim MS, Sloan ED (1987) Theory and measurement of hydrate dissociation. AIChE J 33:747–752

Yoon JH, Kawamura T, Yamamoto Y, Komai T (2004) Transformation of methane hydrate to carbon dioxide hydrate: in situ Raman spectroscopic observations. J Am Chem Soc 108:5057–5059

Yousif MH, Abass HH, Selim MS, Sloan ED (1991) Experimental and theoretical investigation of methane-gas-hydrate dissociation in porous media. SPE Reserv Eng 6:69–76

Index

G

H

I

K

L

M

N

O

P

R

S

T

U

V

W

MIX
Papier aus verantwortungsvollen Quellen
Paper from responsible sources
FSC® C105338

If you have any concerns about our products,
you can contact us on
ProductSafety@springernature.com

In case Publisher is established outside the EU,
the EU authorized representative is:
Springer Nature Customer Service Center GmbH
Europaplatz 3, 69115 Heidelberg, Germany

Printed by Libri Plureos GmbH
in Hamburg, Germany